NURSE ● TEST™
A REVIEW SERIES

S0-AXR-169

Fundamentals of Nursing

SECOND EDITION

Fundamentals of Nursing

SECOND EDITION

Elizabeth (Libby) Ann Archer, RN, MSN

Assistant Professor
Baptist College of Health Sciences
Memphis

Billie E. Ward, RN, MSN

Nursing Instructor
Bishop State Community College
Mobile, Ala.

SPRINGHOUSE CORPORATION ● SPRINGHOUSE, PENNSYLVANIA

STAFF

Vice President
Matthew Cahill

Editorial Director
Darlene Barela Cooke

Clinical Director
Judith Schilling McCann, RN, MSN

Art Director
John Hubbard

Managing Editor
David Moreau

Clinical Project Manager
Beverly Ann Tscheschlog, RN

Editor
Laura Poole

Copy Editors
Brenna H. Mayer (manager), Shana
Harrington, Jake Marcus Cipolla,
Kathryn A. Marino, Pamela Wingrod

Designers
Arlene Putterman (associate art director),
Kate Nichols (book design), Donna S.
Morris (project manager), Stephanie
Peters (cover design), Jon Nelson (cover
illustration)

Manufacturing
Deborah Meiris (director), Patricia K.
Dorshaw (manager), Otto Mezei (book
production manager)

Editorial Assistants
Beverly Lane, Marcia Mills, Liz Schaeffer

The clinical procedures described and recommended in
this publication are based on research and consultation
with nursing, medical, and legal authorities. To the best
of our knowledge, these procedures reflect currently accepted practice; nevertheless, they cannot be considered
absolute and universal recommendations. For individual application, all recommendations must be considered in light of the patient's clinical condition and, before administration of new or infrequently used drugs,
in light of the latest package-insert information. The
author and the publisher disclaim responsibility for any
adverse effects resulting directly or indirectly from the
suggested procedures, from any undetected errors, or
from the reader's misunderstanding of the text.

©2000 by Springhouse Corporation. All rights reserved.
No part of this publication may be used or reproduced
in any manner whatsoever without written permission
except for brief quotations embodied in critical articles
and reviews. For information, write Springhouse
Corporation, 1111 Bethlehem Pike, P.O. Box 908,
Springhouse, PA 19477-0908. Authorization to photocopy items for internal or personal use, or the internal
or personal use of specific clients, is granted by
Springhouse Corporation for users registered with the
Copyright Clearance Center (CCC) Transactional
Reporting Service, provided that the fee of $.75 per
page is paid directly to CCC, 222 Rosewood Dr.,
Danvers, MA 01923. For those organizations that have
been granted a photocopy license by CCC, a separate
system of payment has been arranged. The fee code for
users of the Transactional Reporting Service is
1582550018/2000 $00.00 + $.75.

Printed in the United States of America.

R A member of the Reed Elsevier plc group

NTF2-010899

Library of Congress Cataloging-in-Publication Data
Fundamentals of nursing / [edited by] Elizabeth Archer, Billie E. Ward.
p. cm. — (Nurse test)
Included bibliographical references and index.
1. Nursing Examinations, questions etc. I. Archer, Elizabeth. II. Ward, Billie E. III. Series: NurseTest.
[DNLM: 1. Nursing Care Examination Questions. 2. Nurse-Patient Relations Examination Questions. WY18.2 F981 2000]
RT55.F86 2000
610.73'076—dc21
DNLM/DLC
ISBN 1-58255-001-8 (alk. paper)   99-30117

CONTENTS

CONTRIBUTORS

Janet T. Barrett, RN, BSN, MSN, PhD
Director of BSN Program
Deaconess College of Nursing
St. Louis

Deborah Becker, MSN, CRNP, CCRN
Lecturer — Adult Critical Care Nurse
Practitioner Program
School of Nursing
University of Pennsylvania
Philadelphia

Lori Musolf Neri, RN, MSN, CCRN
Adjunct Clinical Instructor
Villanova (Pa.) University
Staff Nurse
North Penn Hospital
Lansdale, Pa.

Suzanne Omrod-Schmidt, RN, MSN
Lecturer — Adult Health and Illness
Program Assistant
BSN/MSN Direct Entry Program
School of Nursing
University of Pennsylvania
Philadelphia

Bruce Austin Scott, RNCS, MSN
Nursing Instructor
San Joaquin Delta College
Staff Nurse
St. Joseph Medical Center
Stockton, Calif.

Shelley E. Yeager, RN, MSN
Nursing Instructor
Allentown (Pa.) College of St. Francis
de Sales

INTRODUCTION

The commitment to a successful nursing career requires you to demonstrate that you have the knowledge to be a competent nurse. Passing rigorous challenge examinations and the NCLEX licensure examination are examples of your commitment to promoting high-quality nursing care.

NurseTest Fundamentals of Nursing, Second Edition, is an in-depth question-and-answer review book that provides hundreds of multiple-choice questions to test your knowledge of the fundamentals of nursing. Questions are organized into chapters covering medical terms, the nursing process, foundations for nursing practice, communication, physiological needs, psychosocial needs, patients with special needs, and diagnostic tests, therapies, and treatments.

Each question consists of a stem with four answer options. After answering the questions on the answer sheet, you can check your responses against the correct answers provided. When appropriate, the answers include rationales for correct and incorrect options.

As an added feature, each rationale is followed by the corresponding nursing process step, client needs category and subcategory, and taxonomic level for each question, when applicable. Fundamentals of Nursing courses usually cover such topics as nursing history, educational preparation, professional organizations, guidelines for and concepts of professional nursing, and ethical and legal implications of nursing. Because nursing process steps, client needs categories, and client needs subcategories do not always apply to questions on such topics, only the taxonomic level for each question is provided in chapters 2 and 3; for the remaining chapters, the abbreviation N/A is used to designate a step or category that is not applicable to the question.

Nursing process steps

When studying for a fundamentals of nursing examination, keep in mind that the questions are designed to test your knowledge of the five steps in the nursing process: *assessment, analysis, planning, implementation,* and *evaluation.* Each step has a unique function in the delivery of quality nursing care.

A. Safe, effective care environment
  1. *Management of care*
    Providing integrated, cost-effective care to patients by coordinating, supervising, or collaborating with members of the multidisciplinary health care team
  2. *Safety and infection control*
    Protecting patients and health care personnel from environmental hazards

B. Health promotion and maintenance
1. *Growth and development through the life span*
Assisting the patient and significant others through the normal expected stages of growth and development from conception through advanced old age
2. *Prevention and early detection of disease*
Managing and providing care for patients in need of prevention and early detection of health problems

C. Psychosocial integrity
1. *Coping and adaptation*
Promoting patient's ability to cope, adapt, or troubleshoot situations related to illnesses or stressful events
2. *Psychosocial adaptation*
Managing and providing care for patients with acute or chronic mental illnesses

D. Physiological integrity
1. *Basic care and comfort*
Providing comfort and assistance in the performance of activities of daily living
2. *Pharmacological and parenteral therapies*
Managing and providing care related to the administration of medications and parenteral therapies
3. *Reduction of risk potential*
Reducing the likelihood that patients will develop complications or health problems related to existing conditions, treatments, or procedures
4. *Physiological adaptation*
Managing and providing care to patients with acute, chronic, or life-threatening physical health conditions

Taxonomic levels

The taxonomic or cognitive level of questions refers to the type of mental activity required to answer the question as defined in the cognitive domain. The NCLEX examination consists of multiple-choice questions at the cognitive levels of *knowledge, apprehension, application,* and *analysis*. Because the practice of nursing requires application of knowledge, skills, and abilities, most of the questions in the examination are written at the application or analysis level of cognitive ability.

Multiple-choice questions

Multiple-choice questions consist of an introductory statement, the stem, and four options. The introductory statement contains information about the clinical situation, and presents information about the patient, the disease, or interventions being done. The stem is the specific question you are to answer. The four options include, of course, the correct response and three distracters, which are options that are designed to resemble the correct answer but are incorrect. To successfully answer multiple-choice questions, you must separate what the question is telling you from what it's asking. Identify whether the stem is asking for a true response or a false response.

True response stems ask you to identify appropriate nursing actions, prioritize nursing actions, or identify safe nursing judgments or therapeutic nursing responses. True response statements may also ask you to identify the action or statement that demonstrates success with teaching of the patient. Key words used with true response stems include *most, best, initially should, chief, immediate, indicative,* and *understands.*

False response stems are seeking wrong or negative information. Some key terms include *inappropriate, contraindicated, least important, lowest priority, required further instruction, unsafe,* and *least likely.*

Watch for such terms as *early, late, immediate, before, postoperative,* and *on admission.*

Thus, the first step in selecting the correct answer is identifying whether the stem is asking for a true or false response. Then read each option and identify whether it is a true or false response. Eliminate those that do not match what the stem is asking. Pay close attention to the timing indicated in each question. For example:

> According to Maslow's hierarchy of needs, which priority are the individual's esteem needs?
>
> **A.** Second
> **B.** Third
> **C.** Fourth
> **D.** Fifth

Correct answer: **C**

Maslow's hierarchy of needs presents the individual's needs in ascending priority. First-level needs include oxygen, food, and shelter, which are followed by the second-level need of safety. Third-level needs are categorized as love and belonging needs, followed by the fourth level of esteem needs. The final level of needs is self-actualization.

> *Nursing process step:* Assessment
> *Client needs category:* Psychosocial integrity
> *Client needs subcategory:* Coping and adaptation
> *Taxonomic level:* Knowledge

Preparing for the test

Stress and anxiety are normal reactions to testing situations; however, uncontrollable anxiety will hurt your ability to do well. Proper study behaviors and a thorough preparation for an examination are the best ways to reduce the stress experienced before and during an examination. It doesn't matter if you start by reviewing lecture notes, textbooks, or various preparatory NCLEX CAT examinations, you must have a plan. Remember that the studying shouldn't just be for the examination, but to make you a better nurse as you start your career. Begin preparing early and don't try to cram all information into a few nights or hours of studying. The following is a suggested plan for scheduling your study time.

1. Set aside at least 2 to 3 hours for review per session. Review periods of 1 hour or less are often ineffective because of the time spent getting organized.
2. Take short breaks after an hour. Set a stopwatch so that you don't break your concentration by continuously thinking about the time and looking at your watch. When the timer goes off, get up and physically get away from the study area and materials. If you can't trust yourself to keep your break to about 10 minutes, use the timer again. Keep the television off because there are no programs that short!
3. Start to review notes, texts, and other pertinent material.
4. Take a practice test that will diagnose your strengths and weaknesses.
5. One of the best ways to avoid the pitfalls of testing is to place yourself in the environment as often as possible. Practice with other tests as often as possible; compare results and identify your strengths and weaknesses.
6. Doing practice tests is a highly effective way to study and remember the material.

Day of the examination

1. Get up early.
2. Wear comfortable clothes, preferably layers that are easier to adjust to the environment.
3. Eat breakfast.
4. Leave early.
5. Arrive early.
6. Do NOT study while you wait for your turn on the computer.
7. Read, listen to music, relax.
8. Leave notes and books at home.
9. Listen carefully to the instructions given before entering the test room.

Test anxiety

Don't underestimate the power of positive thinking. If you tell yourself often enough "I'm going to fail this test," you just might do so. Therefore, if you keep convincing yourself that you're as prepared as anyone and are going to do well on the test, you're already ahead of the game.

If you find instant retrieval difficult and you feel as if facts and figures are just a jumble in your mind, you'll need to take a moment to relax. The simplest relaxation method is deep breathing. Just lean back in your chair, relax your muscles, and take three very deep breaths (count to 10 while you hold each one). That may be the only relaxing technique you need. Other techniques involve focusing your mind on one thing and excluding everything else. While you're concentrating on one object, your mind can't be thinking of anything else. Thus it will slow down a bit.

Taking the test

You can improve your chances of performing well on multiple-choice tests by learning to choose between the best two options available and by following a set of proven guidelines.

Choosing between the two best options

The first step in choosing between the two best options is to be sure that you understand what the question is asking. Then eliminate the incorrect options. If you're having difficulty choosing between two seemingly correct responses, use the following strategies.

    a. Eliminate similar distracters. If two options are essentially saying the same thing or include the same idea, then neither of them can be the answer. The answer has to be the option that is different.

    b. Reread two seemingly correct options. If two options seem equally correct, reread them carefully; there must be some difference between them. Reread the stem; you may notice something you missed before.

    c. Look for a global response. A more general statement may also include correct ideas from other options.

Guidelines to remember

- Budget your time. Although you may not know exactly how many questions you'll be asked to answer, you can estimate a little over 1 minute per question. Keep moving at a steady pace.
- Read each question thoroughly but quickly. In general, your first reaction to a question is the correct one. Remember that the examination is designed to determine if you're minimally competent and safe.
- Concentrate on one item at a time. Don't worry about how many questions you'll have to answer.
- Answer questions as if the situation were ideal. Assume the nurse had all the time and resources needed. You're only concerned about one patient, the one in the question.
- Focus on the key words in the stem.
- Identify whether the stem is seeking a true response or a false response. Those stems asking for false responses are easily misread.
- Reword a difficult stem.
- Try answering the question before you've read the options provided.
- Always read all options before selecting the best one.

- Relate each option to the stem.
- Use logic and common sense to figure out the correct response.
- Remember that the correct option will tend to have greater applicability and flexibility.
- Clueless? Look for clues in answer choices instead of in the stem of the question.
- Don't compare the situation to what your hospital does or to other experiences you may have had.

JUDITH ANN DRISCOLL, RN, MEd, MSN
Assistant Professor
Deaconess College of Nursing
St. Louis

CHAPTER 1

MEDICAL TERMS AND DEFINITIONS

QUESTIONS

1. The nurse manager has requested that a nurse perform a self-assessment prior to an annual performance evaluation. Using Maslow's model of self-actualization, the nurse understands that self-actualization is at the top of the pyramid and refers to:

    A. becoming the person you would like to be by making the most of your physical, mental, and emotional competencies.

    B. needing recognition, usefulness, independence, dignity, and freedom.

    C. needing safety, security, and protection.

    D. needing to belong to a group.

2. The nurse is evaluating a patient complaining of shortness of breath. The nurse assesses the patient's respiratory rate to be 26 breaths/minute and documents that the patient is tachypneic. The nurse understands that tachypnea means:

    A. frequent bowel sounds.

    B. heart rate greater than 100 beats/minute.

    C. hyperventilation.

    D. respiratory rate greater than 20 breaths/minute.

3. A patient is evaluated at a doctor's office for complaints of insomnia. The nurse understands that a correct definition of insomnia may be:

    A. crawling leg sensations while asleep.

    B. difficulty initiating and maintaining sleep.

    C. episodes of drowsiness followed by brief naps.

    D. sleep apnea.

4. The nurse is assessing the lymphatic system. Which organ is defined as lymphatic tissue?

    A. Pancreas

    B. Sinus

    C. Thymus

    D. Thyroid

5. The nurse is assessing a wound and observes new tissue growing onto the wound from surrounding tissue. The nurse understands that this situation is called:

A. angiogenesis.
B. granulation.
C. normal wound healing.
D. wound contraction.

6. A patient is admitted to the emergency department with complaints of pain to her surgical incision. The doctor diagnoses the patient with wound dehiscence. The nurse understands that wound dehiscence is:

A. a normal postoperative occurrence.
B. a ruptured organ.
C. an opening of the wound edges.
D. the protrusion of internal organs at a wound site.

7. A patient with extensive third-degree burns begins to look poorly. Test results indicate inadequately circulating blood volume due to fluid shifts and a loss of plasma proteins. The nurse alerts the doctor because she suspects:

A. cardiogenic shock.
B. distributive shock.
C. hypovolemic shock.
D. obstructive shock.

8. An oncologist is evaluating a patient diagnosed with breast cancer. Further diagnostic studies are ordered to determine if the cancer has spread. The nurse understands that the ability of cancer to spread to other parts of the body is called:

A. dysplasia.
B. hyperplasia.
C. metaplasia.
D. metastasis.

9. A surgical nurse understands that asepsis must be maintained to limit the occurrence and spread of pathogens. The nurse understands that a pathogen is:

A. a microorganism capable of transmitting a disease.
B. development of a disease.
C. normal flora.
D. a portal of entry.

10. The nurse is assessing a patient admitted to the intensive care unit with complaints of chest pain and shortness of breath. The patient underwent a total hip replacement 3 weeks ago. The nurse understands that an embolism may be:

    A. bacteria, fragments of bone, or clotted blood, fat, or air.
    B. clotted blood, fat, air, or ischemic brain tissue.
    C. fragments of bone or clotted blood, fat, or air.
    D. fragments of clotted blood, tumor, fat, bacteria, or air.

11. The nurse is assessing a patient for meningeal irritation. Upon lifting the patient's head rapidly from the bed, she observes flexion of both thighs at the hip and flexure movements of the ankles and knees. She understands that these signs are characteristically known as:

    A. Brudzinski's sign.
    B. Homans' sign.
    C. Kernig's sign.
    D. Rinne's test.

12. A patient is admitted to the emergency department complaining of weakness. Assessment reveals pallor, diaphoresis, and nausea. Vital signs are within normal ranges. The doctor diagnoses the patient with syncope. The nurse understands that syncope refers to:

    A. dizziness.
    B. fainting.
    C. head injury.
    D. vertigo.

13. A patient complaining of numbness, difficulty in coordination, and loss of balance is admitted to the neurology unit. The nurse understands that these clinical manifestations may be due to a progressive degenerative disease that affects the myelin sheath, leading to various degenerative symptoms. This disease is characteristically known as:

    A. lymphoma.
    B. multiple sclerosis.
    C. myasthenia gravis.
    D. Parkinson's disease.

14. A mother presents her child to the pediatrician's office for evaluation. The child is rubbing his left eye. The nurse's assessment reveals redness to the left sclera. The sclera:

    A. bends light to the retina.
    B. is a fibrous protective covering of the eye.
    C. is a thin pigmented diaphragm.
    D. produces aqueous humor.

15. The nurse assesses a patient complaining of fever, headache, and a dry, irritating cough. The nurse suspects an inflammation in the portion of the respiratory tract that passes from the larynx to the two main bronchi. This tract is known as the:

    A. alveoli.
    B. bronchioles.
    C. main bronchi.
    D. trachea.

16. The nurse assesses a patient complaining of frequent episodes of epistaxis. The nurse knows the patient has:

    A. an enlarged spleen.
    B. a tendency to bruise easily.
    C. nosebleeds.
    D. seizures.

17. The doctor evaluates a patient experiencing prolonged spasms of the airways and orders epinephrine subcutaneously. The nurse administers the medication with the understanding that prolonged spasms of the airways are a symptom best described as:

    A. asthma.
    B. bronchitis.
    C. bronchospasm.
    D. emphysema.

18. The nurse assesses a patient diagnosed with chronic bronchitis. Observation reveals a barrel chest, often seen in patients with chronic bronchitis. A person with a barrel chest has an anterior-to-posterior diameter of:

    A. 1:1.
    B. 1:2.
    C. 1:2 to 5:7.
    D. 5:7.

19. The nurse assesses a patient complaining of chills, fever, and stabbing chest pain. The doctor diagnoses the patient with pneumonia. The nurse knows that pneumonia is:

    A. a form of obstructive lung disease.
    B. an acute viral infection of the respiratory tract.
    C. an inflammatory process of the lungs resulting in increased interstitial and alveolar fluid.
    D. a collapse of lung tissue at the structural level.

20. A patient is admitted to the surgical unit following abdominal surgery. He has an indwelling urinary catheter connected to a drainage unit. Three days after admission, the nurse notes that the patient's urine is cloudy and he has spiked a temperature of 101.2° F (38.4° C). The nurse suspects that the patient may have developed a nosocomial infection. The nurse understands that nosocomial infections are:

    A. caused by normal flora.
    B. acquired in a hospital.
    C. impossible to prevent.
    D. acquired in a home care setting.

21. The doctor orders mechanical ventilation for a patient admitted to the intensive care unit with a diagnosis of ARDS. The nurse assists with placing the patient on a mechanical ventilator with the understanding that ARDS stands for:

    A. acute renal disease syndrome.
    B. acute renal distress syndrome.
    C. adult respiratory disease syndrome.
    D. adult respiratory distress syndrome.

22. A patient is admitted to the emergency department with a stab wound to the chest. The patient has distended neck veins and muffled heart sounds. The nurse immediately notifies the doctor, suspecting these symptoms are caused by a process of fluid accumulation in the pericardial cavity, restricting diastolic filling. This process is known as:

    A. cardiac tamponade.
    B. cardiomyopathy.
    C. heart failure.
    D. pericarditis.

23. The nurse's assessment of a patient's medical history reveals rheumatic fever. The nurse knows that rheumatic fever can result in stenosis of one of the heart structures that maintains one-way flow of blood. These structures within the heart are called:

    A. atria.
    B. chordae tendineae.
    C. valves.
    D. ventricles.

24. The nurse is assessing a patient with a diagnosis of atherosclerosis. The nurse understands that atherosclerosis involves the accumulation of lipids, calcium, and blood components on the intima layer of the artery. The arteries are composed of three layers. The layer that is responsible for constriction and dilation of the artery is:

    A. tunica adventitia.
    B. tunica aorta.
    C. tunica intima.
    D. tunica media.

25. The doctor orders intra-arterial monitoring to obtain continuous blood pressure in a critically ill patient. Collateral circulation should be assessed prior to selection of an arterial site. The test used to assess the patency of the radial and ulnar arteries is the:

    A. Allen's test.
    B. Homans' sign.
    C. Trendelenburg's test.
    D. Weber's test.

26. A patient has been diagnosed with thrombocytopenia. The nurse understands that thrombocytopenia can result from decreased production of platelets by the bone marrow. The nurse knows that platelets are also known as:

    A. agranulocytes.
    B. granulocytes.
    C. leukocytes.
    D. thrombocytes.

27. The nurse is caring for a patient with a diagnosis of hypertension. Diagnostic laboratory studies indicate an increase in the number of red blood cells. The nurse understands that an increase in the production of red blood cells is known as:

    A. fibrinolysis.
    B. hematopoiesis.
    C. hemophilia.
    D. polycythemia.

28. The nurse is assessing a patient complaining of weakness and fatigue. Diagnostic laboratory studies indicate a decrease in vitamin B_{12} absorption. The nurse understands that anemia related to a decrease in vitamin B_{12} absorption is:

    A. hemolytic anemia.
    B. iron deficiency anemia.
    C. megaloblastic anemia.
    D. pernicious anemia.

29. A patient is admitted to the medical unit with a diagnosis of hepatitis. Assessment reveals jaundice. The nurse knows that jaundice is:

    A. an accumulation of bilirubin in the gallbladder.
    B. an accumulation of bilirubin in the blood.
    C. hyperactivity of macrophages within the liver.
    D. hyperactivity of macrophages within the spleen.

30. A patient is bleeding from mucous membranes and the GI and urinary tracts. Laboratory diagnostic tests exhibit a prolonged prothrombin time and a reduced platelet count. The nurse understands that the pathological condition caused by a diffuse coagulation is known as:

    A. anemia.
    B. disseminated intravascular coagulation.
    C. hemophilia.
    D. purpura.

31. The doctor schedules a patient for diagnostic studies based on the patient's complaint of right flank pain and blood-tinged urine. When ordering the diagnostic studies, the nurse understands that the injection of I.V. dye that is filtered through the kidneys and excreted through the urinary tact is called:

    A. cystourethrography.
    B. excretory urography.
    C. KUB.
    D. renal phlebography.

32. A patient has an indwelling urinary catheter attached to drainage. When transporting the patient to another area, the nurse hooks the drainage bag below the level of the bladder to prevent a backward flow of urine. A backward flow of urine is known as:

    A. oliguria.
    B. urinary incontinence.
    C. urinary reflux.
    D. urine retention.

33. A patient complains of chills, fever, and flank pain. The doctor diagnoses the patient with pyelonephritis. The nurse understands that pyelonephritis is:

    A. a bacterial infection causing inflammation of the renal pelvis.
    B. an abrupt loss of kidney function.
    C. an immunologic reaction resulting in changes in the glomerular structure.
    D. irreversible renal failure.

34. While examining a patient's oral cavity, the nurse observes redness and excoriation at the anterior portion of the roof of the mouth. This area is known as the:

    A. buccal cavity.
    B. hard palate.
    C. oral cavity.
    D. soft palate.

35. The nurse is performing a sitz bath for a patient with a diagnosis of perianal varicose veins. The nurse understands that perianal varicose veins are also known as:

    A. abscesses.
    B. cysts.
    C. fistulas.
    D. hemorrhoids.

36. The nurse is caring for a patient who is profusely vomiting bright red blood. The doctor orders a Minnesota tube (or Sengstaken-Blakemore tube). The purpose of this tube is to:

    A. collect liquid stool.
    B. drain peritoneal abscesses.
    C. provide enteral feedings.
    D. stop bleeding of esophageal varices.

37. A patient has been diagnosed with hepatic encephalopathy. The nurse observes flapping tremors. The nurse understands that flapping tremors associated with hepatic encephalopathy are also known as:

    A. aphasia.
    B. ascites.
    C. astasia.
    D. asterixis.

SITUATION: *Samuel Fenstermacher is admitted to the medical-surgical unit with a diagnosis of cerebral vascular accident. The nurse's assessment reveals hemiplegia and slurred speech.*

Questions 38 and 39 relate to this situation.

38. Hemiplegia refers to:

    A. paralysis of one side of the body.
    B. paralysis of the upper or lower half of the body.
    C. paresthesia of one side of the body.
    D. paresthesia of the upper or lower half of the body.

39. A language defect caused by a cerebral cortex disorder that can affect talking, writing, reading, and understanding words is:

    A. agnosia.
    B. aphasia.
    C. apraxia.
    D. dysarthria.

SITUATION: *Becky Del Sordo is admitted to the emergency department with a history of tonic-clonic seizures. She's currently experiencing seizure activity, which the family states began at home. The nurse's assessment determines that the patient is currently in the clonic phase of the seizure. The doctor diagnoses Ms. Del Sordo with status epilepticus and orders phenytoin (Dilantin) to be given I.V.*

Questions 40 to 42 relate to this situation.

40. Tonic-clonic seizures are also referred to as:

    A. absence seizures.
    B. grand mal seizures.
    C. myoclonic seizures.
    D. petit mal seizures.

41. The clonic phase of a seizure includes which of the following symptoms?

    A. Relaxation of muscles, periods of apnea, incontinence, and excessive saliva
    B. Rhythmic jerking, relaxation of muscles, incontinence, and excessive saliva
    C. Rhythmic jerking, stiffening of entire body, and incontinence
    D. Stiffening of entire body, periods of apnea, and clenched fists

42. Status epilepticus is a:

    A. reaction to seizure medications.
    B. state of continuous or successive seizures.
    C. state of continuous petit mal seizures.
    D. term used only for seizures that occur in succession.

SITUATION: *Herbert Heller is admitted with a medical diagnosis of emphysema. The nurse's assessment reveals clubbing of the nail beds and a musical, hissing sound that is heard on auscultation of the lungs.*

Questions 43 and 44 relate to this situation.

43. By definition, clubbing of the nail beds occurs when their angle increases. This angle would be greater than:

A. 120 degrees.
B. 160 degrees.
C. 180 degrees.
D. 200 degrees.

44. The nurse listens to Mr. Heller's lungs and notes a hissing or musical sound, such as air flowing through a narrow passage. The nurse documents hearing:

A. crackles.
B. normal breath sounds.
C. wheezes.
D. rhonchi.

SITUATION: *John Cooper has been admitted to the cardiac intensive care unit with complaints of chest pain and shortness of breath. He stated that he took nitroglycerin sublingually prior to admission to the hospital, but his pain wasn't relieved. Mr. Cooper is placed on a cardiac monitor, and the nurse observes a ventricular arrhythmia.*

Questions 45 to 48 relate to this situation.

45. The largest and most muscular chamber of the heart is the:

A. left atrium.
B. left ventricle.
C. right atrium.
D. right ventricle.

46. An arrhythmia is:

B. a fast and irregular heart rhythm.
B. a fast heart rhythm.
C. an abnormal heart rhythm.
D. difficulty breathing.

47. Cardiac output is defined as:

A. blood pressure plus heart rate.
B. blood pressure times heart rate.
C. stroke volume plus heart rate.
D. stroke volume times heart rate.

48. To administer nitroglycerin sublingually, the nurse:

A. inserts it rectally.
B. places it in the buccal area.
C. places it on the chest wall.
D. places it under the tongue.

SITUATION: *Sheila Killian has a history of hypertension and complains of short-ness of breath when she lies flat. The nurse assesses her blood pressure to have a dias-tolic reading of 200 mm Hg.*

Questions 49 and 50 relate to this situation.

49. The nurse documents that Mrs. Killian has:

    A. dyspnea.
    B. exertional dyspnea.
    C. orthopnea.
    D. paroxysmal nocturnal dyspnea.

50. Diastole is defined as:

    A. a measure of contractility.
    B. the amount of blood ejected from the ventricles with each contraction of the heart.
    C. the period when the heart contracts and the muscle fibers are tight and short.
    D. the period when the muscle fibers are stretched and the heart's cavities fill with blood.

5. *Correct answer:* **B**

When new tissue grows onto the wound from surrounding tissue, this is called granulation. Although granulation is part of wound healing, it isn't the entire process. Wound contraction occurs when the edges of the wound come together. Angiogenesis is the development of new blood vessels in the wound bed.

> *Nursing process step:* Assessment
> *Client needs category:* Physiological integrity
> *Client needs subcategory:* Reduction of risk potential
> *Taxonomic level:* Knowledge

6. *Correct answer:* **C**

Wound dehiscence is an opening of the wound edges. Evisceration is the protrusion of internal organs at a wound site. An opening of the wound edges is not a normal postoperative occurrence. A ruptured organ is not the definition of a wound dehiscence.

> *Nursing process step:* Assessment
> *Client needs category:* Physiological integrity
> *Client needs subcategory:* Physiological adaptation
> *Taxonomic level:* Knowledge

7. *Correct answer:* **C**

Burns are a cause of hypovolemic shock. Cardiogenic shock occurs when there is an inadequate pumping action of the heart. Distributive shock is due to changes in blood vessel tone. Obstructive shock is a fictional term.

> *Nursing process step:* Implementation
> *Client needs category:* Psychological integrity
> *Client needs subcategory:* Physiological adaptation
> *Taxonomic level:* Comprehension

8. *Correct answer:* **D**

The ability of cancer cells to spread is known as metastasis. Dysplasia refers to an alteration in size, shape, or arrangement of differentiated cells. Hyperplasia refers to an increase in the normal number of cells. Metaplasia refers to replacement of one differentiated cell by another.

> *Nursing process step:* Assessment
> *Client needs category:* Physiological integrity
> *Client needs subcategory:* Physiological adaptation
> *Taxonomic level:* Knowledge

9. *Correct answer:* **A**

A pathogen is a microorganism capable of transmitting a disease, not normal flora or a portal of entry. Development of a disease is pathogenesis.

> *Nursing process step:* Implementation
> *Client needs category:* Safe, effective care environment
> *Client needs subcategory:* Safety and infection control
> *Taxonomic level:* Knowledge

10. *Correct answer:* **D**

By definition, an embolism may be fragments of clotted blood, tumor, fat, bacteria, or air. Any of these particles can form an embolism. Bone and ischemic brain tissue can't form an embolism.

> *Nursing process step:* Assessment
> *Client needs category:* Psychological integrity
> *Client needs subcategory:* Physiological adaptation
> *Taxonomic level:* Knowledge

11. *Correct answer:* **A**

Flexion of both thighs at the hip and flexure movements of the ankles and knees indicate Brudzinski's sign. Kernig's sign assesses meningeal irritation by flexing the knee at a 90-degree angle, and spasm occurs at the hamstring muscle. Homans' sign tests for deep vein thrombosis by applying gentle pressure on the calf muscle. Rinne's test checks for air and bone conduction related to hearing using a tuning fork.

> *Nursing process step:* Assessment
> *Client needs category:* Physiological integrity
> *Client needs subcategory:* Physiological adaptation
> *Taxonomic level:* Comprehension

12. *Correct answer:* **B**

Syncope refers to fainting, not dizziness or vertigo. A head injury can cause syncope but doesn't define it.

> *Nursing process step:* Assessment
> *Client needs category:* Physiological integrity
> *Client needs subcategory:* Physiological adaptation
> *Taxonomic level:* Knowledge

13. *Correct answer:* **B**

Multiple sclerosis is a degenerative disease of the myelin sheath in the brain and spinal cord resulting in muscle tremors and partial or total paralysis. Parkinson's disease causes degeneration of dopamine-producing cells, leading to tremors and other symptoms. Myasthenia gravis is an autoimmune disease producing weakness. Lymphoma is cancer of the lymphatic system.

Nursing process step: Assessment
Client needs category: Physiological integrity
Client needs subcategory: Physiological adaptation
Taxonomic level: Comprehension

14. *Correct answer:* **B**

The sclera is a fibrous protective covering of the eye. The cornea bends light to the retina. The iris is a thin, pigmented diaphragm. The ciliary body produces aqueous humor.

Nursing process step: Assessment
Client needs category: Physiological integrity
Client needs subcategory: Physiological adaptation
Taxonomic level: Knowledge

15. *Correct answer:* **D**

The trachea is the respiratory tract that passes from the larynx to the two main bronchi. Alveoli are terminal ends of the bronchi, bronchioles are small bronchi, and the main bronchi is the right main stem bronchus.

Nursing process step: Assessment
Client needs category: Physiological integrity
Client needs subcategory: Physiological adaptation
Taxonomic level: Knowledge

16. *Correct answer:* **C**

Epistaxis refers to nosebleeds. The other phrases don't correctly define this term.

Nursing process step: Assessment
Client needs category: Physiological integrity
Client needs subcategory: Physiological adaptation
Taxonomic level: Knowledge

17. *Correct answer:* **C**

Bronchospasm is prolonged spasms of the airways. Asthma is a disease characterized by periods of bronchospasm. Bronchitis and emphysema are chronic obstructive pulmonary diseases.

Nursing process step: Implementation
Client needs category: Physiological integrity
Client needs subcategory: Pharmacological and parenteral therapies
Taxonomic level: Comprehension

18. *Correct answer:* **A**

A barrel chest is 1:1. Ratios of 1:2 to 5:7 are all normal chest diameters.

> *Nursing process step:* Assessment
> *Client needs category:* Physiological integrity
> *Client needs subcategory:* Physiological adaptation
> *Taxonomic level:* Knowledge

19. *Correct answer:* **C**

Pneumonia is an inflammatory process of the lungs resulting in increased interstitial and alveolar fluid. Many lung diseases, such as bronchitis and emphysema, are classified as obstructive. An acute viral infection of the respiratory tract is influenza. A collapse of lung tissue at the structural level is atelectasis.

> *Nursing process step:* Assessment
> *Client needs category:* Physiological integrity
> *Client needs subcategory:* Physiological adaptation
> *Taxonomic level:* Knowledge

20. *Correct answer:* **B**

Nosocomial infections are acquired in a hospital. They aren't caused by normal flora or acquired in a home care setting, and they aren't impossible to prevent. Hand washing and aseptic techniques are two ways nosocomial infections can be prevented.

> *Nursing process step:* Assessment
> *Client needs category:* Safe, effective care environment
> *Client needs subcategory:* Safety and infection control
> *Taxonomic level:* Knowledge

21. *Correct answer:* **D**

ARDS stands for adult respiratory distress syndrome. The other answers are fictitious definitions.

> *Nursing process step:* Implementation
> *Client needs category:* Physiological integrity
> *Client needs subcategory:* Physiological adaptation
> *Taxonomic level:* Knowledge

22. *Correct answer:* **A**

Accumulation of fluid in the pericardial cavity, restricting diastolic filling, is known as cardiac tamponade. Cardiomyopathy is a disease of the myocardium. Heart failure is failure of the heart's pumping action caused by either or both pulmonary and peripheral effects. Pericarditis is inflammation of the parietal and visceral pericardium.

Nursing process step: Implementation
Client needs category: Physiological integrity
Client needs subcategory: Physiological adaptation
Taxonomic level: Comprehension

23. *Correct answer:* **C**

The four valves — aortic, pulmonic, tricuspid, and mitral — allow for unilateral flow of blood. Chordae tendineae hold the valves in place. The atria are holding chambers for blood within the heart. The ventricles act as pumps, pushing blood through the heart.

Nursing process step: Assessment
Client needs category: Physiological integrity
Client needs subcategory: Physiological adaptation
Taxonomic level: Knowledge

24. *Correct answer:* **D**

The tunica media controls the diameter of a blood vessel by constricting and dilating. The tunica adventitia is the outermost layer, which gives arteries their shape and support. Tunica aorta is a fictional term. Tunica intima is the innermost portion of the vessel, where the blood flows.

Nursing process step: Assessment
Client needs category: Physiological integrity
Client needs subcategory: Physiological adaptation
Taxonomic level: Knowledge

25. *Correct answer:* **A**

Allen's test determines patency by compressing the arteries and releasing them one at a time. The hand should regain color within 6 seconds. Homans' sign tests for blood clots in the leg. Trendelenburg's test checks for incompetent valves in the legs. Weber's test determines lateralization of bone conduction by using a tuning fork.

Nursing process step: Implementation
Client needs category: Physiological integrity
Client needs subcategory: Reduction of risk potential
Taxonomic level: Comprehension

26. *Correct answer:* **D**

Platelets are also called thrombocytes. Leukocytes are another name for white blood cells. Agranulocytes and granulocytes are two types of white blood cells.

Nursing process step: Assessment
Client needs category: Psychological integrity
Client needs subcategory: Psychological adaptation
Taxonomic level: Knowledge

27. *Correct answer:* **D**

Polycythemia is an increase in the production of red blood cells. Fibrinolysis is destruction of a clot. Hemophilia is a disorder of the coagulation system. Hematopoiesis is the formation of blood from the bone marrow.

Nursing process step: Assessment
Client needs category: Physiological integrity
Client needs subcategory: Physiological adaptation
Taxonomic level: Knowledge

28. *Correct answer:* **D**

Pernicious anemia is a decrease in vitamin B_{12} absorption. Megaloblastic anemia is a deficiency in B_{12} and folic acid. Hemolytic anemia is caused by the shortened life of red blood cells, an increase in the number of red blood cells being destroyed, and a failure of the bone marrow to replace red blood cells. Iron deficiency anemia is due to inadequate absorption, or loss, of iron.

Nursing process step: Assessment
Client needs category: Physiological integrity
Client needs subcategory: Psychological adaptation
Taxonomic level: Knowledge

29. *Correct answer:* **B**

Jaundice is an accumulation of bilirubin in the blood. An accumulation of bilirubin in the gallbladder is cholelithiasis. Hyperactivity of macrophages within the liver is hepatomegaly. Hyperactivity of macrophages within the spleen is splenomegaly.

Nursing process step: Assessment
Client needs category: Physiological integrity
Client needs subcategory: Psychological adaptation
Taxonomic level: Knowledge

30. *Correct answer:* **B**

Disseminated intravascular coagulation is pathologic diffuse coagulation. Anemia is a decrease in red blood cells. Hemophilia causes prolonged coagulation time. Purpura is an extravasation of small amounts of blood into mucous membranes and tissue.

Nursing process step: Assessment
Client needs category: Physiological integrity
Client needs subcategory: Psychological adaptation
Taxonomic level: Comprehension

31. *Correct answer:* **B**

A study that consists of the injection of I.V. dye that is filtered through the kidneys and excreted through the urinary tract is called excretory urography. A KUB is an X-ray of the kidneys, ureters, and bladder. A cystourethrography is an X-ray examination of the bladder and urethra. A renal phlebography is a study of the venous system of the kidneys.

Nursing process step: Implementation
Client needs category: Psychological integrity
Client needs subcategory: Psychological adaptation
Taxonomic level: Knowledge

32. *Correct answer:* **C**

A backward flow of urine is known as urinary reflux. Oliguria is diminished urination. Urinary incontinence is an involuntary loss of urine. Urine retention is the inability to expel urine from the bladder.

Nursing process step: Implementation
Client needs category: Physiological integrity
Client needs subcategory: Reduction of risk potential
Taxonomic level: Knowledge

33. *Correct answer:* **A**

Pyelonephritis is a bacterial infection causing inflammation of the renal pelvis. An abrupt loss of kidney function is acute renal failure. Chronic renal failure is irreversible renal failure. An immunologic reaction resulting in changes in the glomerular structure is acute glomerulonephritis.

Nursing process step: Assessment
Client needs category: Physiological integrity
Client needs subcategory: Physiological adaptation
Taxonomic level: Comprehension

34. *Correct answer:* **B**

The anterior portion of the roof of the mouth is known as the hard palate. The posterior portion is the soft palate. The buccal and oral cavities are other terms for the mouth itself.

Nursing process step: Assessment
Client needs category: Health maintenance and promotion
Client needs subcategory: Prevention and detection of early disease
Taxonomic level: Knowledge

35. *Correct answer:* **D**

Perianal varicose veins are hemorrhoids. Abscesses form from an infection by a microorganism. Cysts are closed sacs that may be solid or contain fluid. Fistulas are sinus tracts that occur between two cavities.

Nursing process step: Implementation
Client needs category: Physiological integrity
Client needs subcategory: Physiological adaptation
Taxonomic level: Knowledge

36. *Correct answer:* **D**

A Minnesota tube (or Sengstaken-Blakemore tube) is used to stop bleeding of esophageal varices. This tube isn't used for the other interventions listed.

Nursing process step: Implementation
Client needs category: Physiological integrity
Client needs subcategory: Physiological adaptation
Taxonomic level: Knowledge

37. *Correct answer:* **D**

Flapping tremors associated with hepatic encephalopathy are asterixis. Aphasia is the inability to speak. Ascites is an accumulation of fluid in the peritoneal cavity. Astasia is the inability to stand or sit erect.

Nursing process step: Assessment
Client needs category: Physiological integrity
Client needs subcategory: Physiological adaptation
Taxonomic level: Knowledge

38. *Correct answer:* **A**

Hemiplegia refers to paralysis of one side of the body. Hemiplegia isn't paralysis of the upper or lower half of the body, paresthesia of one side of the body, or paresthesia of the upper or lower half of the body.

Nursing process step: Assessment
Client needs category: Physiological integrity
Client needs subcategory: Physiological adaptation
Taxonomic level: Knowledge

39. *Correct answer:* **B** ˙

Aphasia is a language defect that is caused by a cerebral cortex disorder. Agnosia refers to a disorder in interpreting stimuli. Apraxia refers to the ability to move a body part but the inability to use it purposefully. Dysarthria is difficulty speaking.

> *Nursing process step:* Assessment
> *Client needs category:* Physiological integrity
> *Client needs subcategory:* Physiological adaptation
> *Taxonomic level:* Knowledge

40. *Correct answer:* **B**

Tonic-clonic seizures are also known as grand mal seizures. Absence seizures are another name for petit mal seizures. Myoclonic seizures are classified as minor motor seizures.

> *Nursing process step:* Assessment
> *Client needs category:* Physiological integrity
> *Client needs subcategory:* Physiological adaptation
> *Taxonomic level:* Knowledge

41. *Correct answer:* **B**

The clonic phase of a seizure includes rhythmic jerking, relaxation of muscles, incontinence, and excessive saliva. The other combinations of symptoms listed are from both phases of a seizure.

> *Nursing process step:* Assessment
> *Client needs category:* Physiological integrity
> *Client needs subcategory:* Physiological adaptation
> *Taxonomic level:* Knowledge

42. *Correct answer:* **B**

Status epilepticus is a state of continuous or successive seizures. The other responses do not describe status epilepticus.

> *Nursing process step:* Assessment
> *Client needs category:* Physiological integrity
> *Client needs subcategory:* Physiological adaptation
> *Taxonomic level:* Knowledge

43. *Correct answer:* **B**

Clubbing of the nail beds occurs when their angle increases to greater than 160 degrees.

Nursing process step: Assessment
Client needs category: Physiological integrity
Client needs subcategory: Physiological adaptation
Taxonomic level: Knowledge

44. *Correct answer:* **C**

Wheezes are indicated by a hissing or musical sound, such as air through a narrow passage. They are adventitious lung sounds. Crackles are usually "wet-sounding" breath sounds. Rhonchi are usually coarse breath sounds.

Nursing process step: Assessment
Client needs category: Physiological integrity
Client needs subcategory: Physiological adaptation
Taxonomic level: Knowledge

45. *Correct answer:* **B**

The left ventricle is larger, stronger, and more muscular than the other pumping chamber, the right ventricle. The right and left atria are considered to be blood reservoirs.

Nursing process step: Assessment
Client needs category: Physiological integrity
Client needs subcategory: Physiological adaptation
Taxonomic level: Knowledge

46. *Correct answer:* **C**

Arrhythmia is a broad term for abnormal heart rhythms. A fast heart rhythm may be tachycardia, atrial fibrillation, or ventricular fibrillation. A fast and irregular heart rhythm may be atrial fibrillation. Dyspnea is difficulty breathing.

Nursing process step: Assessment
Client needs category: Physiological integrity
Client needs subcategory: Physiological adaptation
Taxonomic level: Knowledge

47. *Correct answer:* **D**

Cardiac output is defined as stroke volume times heart rate. The other options are incorrect.

Nursing process step: Assessment
Client needs category: Physiological integrity
Client needs subcategory: Physiological adaptation
Taxonomic level: Knowledge

48. *Correct answer:* **D**

Sublingual drugs are administered under the tongue. Administrations of drugs to the chest wall, buccal area, or rectum are usually ordered as such.

> *Nursing process step:* Implementation
> *Client needs category:* Physiological integrity
> *Client needs subcategory:* Pharmacological and parenteral therapies
> *Taxonomic level:* Knowledge

49. *Correct answer:* **C**

Difficulty breathing when lying flat is orthopnea. Dyspnea is generalized shortness of breath. Exertional dyspnea is shortness of breath upon exertion. Paroxysmal nocturnal dyspnea occurs when a patient awakens from sleep due to shortness of breath.

> *Nursing process step:* Assessment
> *Client needs category:* Physiological integrity
> *Client needs subcategory:* Physiological adaptation
> *Taxonomic level:* Knowledge

50. *Correct answer:* **D**

Diastole is the period when the muscle fibers are stretched and the heart's cavities fill with blood. Option A defines inotropic state, option B defines stroke volume, and option C defines systole.

> *Nursing process step:* Assessment
> *Client needs category:* Physiological integrity
> *Client needs subcategory:* Physiological adaptation
> *Taxonomic level:* Knowledge

CHAPTER 2

THE NURSING PROCESS

QUESTIONS

1. What is a characteristic of the nursing process?

    A. Asystematic
    B. Goal-oriented
    C. Inflexible
    D. Stagnant

2. What is the order of the nursing process?

    A. Assessing, diagnosing, implementing, evaluating, planning
    B. Assessing, diagnosing, planning, implementing, evaluating
    C. Diagnosing, assessing, planning, implementing, evaluating
    D. Planning, diagnosing, implementing, assessing, evaluating

3. Which approach to problem solving tests any number of solutions until one is found that works for that particular problem?

    A. Intuition
    B. Routine
    C. The scientific method
    D. Trial and error

4. Which method of reasoning moves from general principles to the collection of specific data or information that confirm or negate a hypothesis?

    A. Deductive
    B. Inductive
    C. Insightful
    D. Rational

5. During the planning phase of the nursing process, which of the following is the "product" developed?

    A. Nursing care plan
    B. Nursing diagnoses
    C. Nursing history
    D. Nursing notes

6. Which range of applications does the nursing process have?

    A. Broad
    B. Distinct
    C. Exact
    D. Narrow

7. Objective data are also known as:

    A. covert data.
    B. inferences.
    C. overt data.
    D. symptoms.

8. Data or information obtained from the assessment of a patient is primarily used by the nurse to:

    A. ascertain the patient's responses to health problems.
    B. assist in constructing the taxonomy of nursing interventions.
    C. determine the effectiveness of the doctor's orders.
    D. identify the patient's disease process.

9. The primary source of data collection in the assessment phase of the nursing process is the:

    A. chart.
    B. patient.
    C. doctor.
    D. family.

10. What is an example of subjective data?

    A. Color of wound drainage
    B. Odor of breath
    C. Respirations of 14 breaths/minute
    D. The patient's statement of "I feel sick to my stomach"

11. Which statement is a difference between comprehensive and focused assessments?

    A. Comprehensive assessments can't include any focused assessments.
    B. Focused assessments are more important than comprehensive assessments.
    C. Focused assessments are usually ongoing and concerning specific problems.
    D. Objective data are included only in comprehensive assessments.

12. Two-year-old Jason's mother states, "Jason vomited 8 ounces of his formula this morning." This statement is an example of:

    A. objective data from a primary source.
    B. objective data from a secondary source.
    C. subjective data from a primary source.
    D. subjective data from a secondary source.

13. The nurse performs a neurologic exam on a patient. After the exam, which of the following should be recorded as objective data?

   A. +4 patellar reflexes in both of the patient's legs
   B. Patient's description of ringing in his ears
   C. Patient's sensations of numbness in his right arm
   D. Patient's statement, "The room is spinning"

14. Which finding obtained during an assessment is considered significant enough to require immediate communication to another member of the health care team?

   A. Change in a patient's heart rate from 72 to 80
   B. Diminished breath sounds in a patient with previously normal breath sounds
   C. Relief noted by a patient from prescribed nausea medication
   D. Weight loss of 2 lb (1 kg) in a 115-lb (52-kg) female patient

15. Impaired gas exchange is listed in the NANDA taxonomy under which of the following human response patterns?

   A. Choosing
   B. Communicating
   C. Exchanging
   D. Relating

16. It is most important to identify the etiology (risk factors) of a nursing diagnosis because doing so:

   A. assists in organizing nursing care of patients with a similar diagnosis.
   B. describes the patient's health problem or response in a few words.
   C. gives direction to the required nursing interventions for the patient.
   D. indicates the presence of a particular health problem in a patient.

17. The nurse determines that her patient's unwillingness to look at her mastectomy scar, her refusal to see visitors, and her statement, "I'm not the same woman anymore" seem to point to a disturbance in the patient's body image. This observation is an example of:

   A. determining strengths and weaknesses.
   B. differentiating nursing from medical diagnoses.
   C. identifying gaps and inconsistencies in data.
   D. recognizing a cluster of significant data.

18. Mr. Bradley, a Jehovah's Witness, refused any blood transfusions during his surgery, became anemic after his operation, and is now tired. Mr. Bradley is more than likely experiencing:

    A. a potential (risk) nursing diagnosis.
    B. a possible nursing diagnosis.
    C. a wellness diagnosis.
    D. an actual nursing diagnosis.

19. The nurse determines that Mrs. Turner's heart rate of 60 beats/minute means that she's bradycardic. This determination is an example of:

    A. a cue.
    B. an inference.
    C. data clustering.
    D. data synthesis.

20. Using Maslow's hierarchy of basic human needs, which of the following nursing diagnoses has the highest priority?

    A. Anxiety related to impending surgery, as evidenced by insomnia
    B. Impaired verbal communication related to tracheostomy, as evidenced by inability to speak
    C. Ineffective breathing pattern related to pain, as evidenced by shortness of breath
    D. Risk for injury related to autoimmune dysfunction

21. When examining a patient's eyes, the nurse considers the patient's age. This action is an example of:

    A. clustering data.
    B. comparing data against standards and norms.
    C. determining gaps in the data.
    D. differentiating cues and inferences.

22. Formulating a nursing diagnosis is a joint function of:

    A. patient and family.
    B. nurse and patient.
    C. nurse and doctor.
    D. doctor and patient.

23. Which of the following is a nursing diagnosis?

    A. Anxiety
    B. Diabetes
    C. Heart failure
    D. Myocardial infarction

24. The nurse sees a patient crying and determines from that observation that the patient is experiencing the nursing diagnosis of dysfunctional grieving. If the nurse has made a diagnostic error, it would most likely be due to which of the following reasons?

    A. Identifying with the patient
    B. Generalizing from experience
    C. Lack of clinical experience
    D. Premature closure

25. Which statement indicates the most appropriate nursing intervention to determine whether the goal of "The patient will demonstrate fluid balance, as evidenced by total fluid intake equals total fluid output" has been met?

    A. Determine the patient's fluid preferences by the end of the shift.
    B. Instruct the patient to drink 1,500 ml of fluid every day.
    C. Measure and record the patient's total fluid intake and output every shift.
    D. Measure the specific gravity of the patient's urine every shift.

26. What is a disadvantage of standardized nursing care plans?

    A. A nurse who uses a standardized nursing care plan may find it very time consuming.
    B. A standardized nursing care plan may not be consistent with acceptable standards of care.
    C. Consent of the patient is necessary to use a standardized nursing care plan.
    D. The unique needs of the patient may not be addressed in a standardized nursing care plan.

27. Which expected outcome is correctly written?

    A. "The patient will be less edematous [swollen] in 24 hours."
    B. "The patient will drink an adequate amount of fluid daily."
    C. "The patient will identify all high-salt foods from a prepared list by discharge."
    D. "The patient will soon sleep well through the night."

28. In the planning phase of the nursing process, the nurse:

    A. analyzes patient data.
    B. carries out nursing interventions.
    C. formulates a nursing diagnosis.
    D. identifies patient goals .

29. When should discharge planning commence?

    A. 24 hours after discharge
    B. The day before discharge
    C. Upon admission
    D. When the patient desires

30. Which patient would benefit from a short-term goal?

    A. Bill, who is having an appendectomy today
    B. Mary, who has extensive burns over 50% of her body
    C. Susan, who has suffered a severe spinal cord injury
    D. Todd, who has heart failure and is awaiting a transplant

31. The scientific reason for selecting a specific nursing intervention supported by clinical research is called a:

    A. criterion.
    B. rationale.
    C. strategy.
    D. theory.

32. An expected outcome on a patient's nursing care plan reads: "Patient will be able to transfer from the bed to a wheelchair without assistance by the end of the week." When the nurse evaluated the patient's progress, the patient was able to transfer from the bed to a wheelchair to go to the physical therapy department without any help from the nurse. Which of the following would be an appropriate evaluative statement for the nurse to place on the patient's nursing care plan?

    A. "Patient was able to transfer from the bed to a wheelchair without assistance."
    B. "Goal impossible to measure."
    C. "Goal met; patient was able to transfer from the bed to a wheelchair without assistance."
    D. "Goal not met."

33. The nurse manager evaluates the new I.V. system instituted in the facility to ascertain whether it has resulted in a decreased incidence of phlebitis in patients with I.V. lines. This is an example of what type of evaluation?

    A. Outcome
    B. Peer
    C. Process
    D. Structure

34. In the evaluating phase of the nursing process, the nurse does which of the following tasks?

    A. Identifies available resources
    B. Carries out nursing interventions
    C. Formulates nursing diagnoses
    D. Measures goal achievement

35. An expected outcome on a patient's care plan reads: "Patient will state seven warning signals of cancer by discharge." When the nurse evaluates the patient's progress, the patient is able to state that a change in a wart or a mole, a sore that doesn't heal, and a change in bowel or bladder habits are warning signals of cancer. Which of the following would be an appropriate evaluative statement for the nurse to place on the patient's nursing care plan?

    A. "Patient understands the warning signals of cancer."
    B. "Goal met; patient cited a change in a wart or a mole, a sore that doesn't heal, and a change in bowel or bladder habits as warning signals of cancer."
    C. "Goal not met."
    D. "Goal partially met; patient able to state only three warning signals of cancer."

36. The goal or expected outcome "Patient will maintain current weight of 165 pounds" can best be evaluated by which of the following measures?

    A. Determining the patient's food preferences
    B. Monitoring dietary intake for each meal
    C. Restricting high-calorie food
    D. Weighing the patient on the same scale

37. A quality assurance nurse sends a questionnaire to patients after discharge to determine their level of satisfaction with the nursing care they received in the facility. What type of nursing audit is this?

    A. Concurrent
    B. Outcome
    C. Retrospective
    D. Terminal

38. The nurse makes the following entry in the patient's record: "Goal not met; patient refuses to attend smoking cessation classes." Because this goal hasn't been met, the nurse should:

    A. develop a completely new nursing care plan.
    B. assign the patient to a more experienced nurse.
    C. critique the steps involved in the development of the goal.
    D. transfer the patient to another facility.

39. The nurse manager evaluates the quality of the nursing care plans developed for the patients on the nursing unit. This is an example of which type of evaluation?

    A. Outcome
    B. Peer
    C. Process
    D. Structure

40. Which statement regarding quality improvement is true?

    A. Inherent in quality improvement is the notion that some processes in a facility can't be improved.
    B. Quality improvement is a continuous process used to evaluate the needs and expectations of customers.
    C. Quality improvement establishes criteria used to recommend punishment for identified deficiencies.
    D. The focus of quality improvement is primarily on the nursing and medical staffs of a facility.

41. In the implementing phase of the nursing process, the nurse does which of the following tasks?

    A. Determines the patient's health status
    B. Identifies available resources
    C. Measures goal achievement
    D. Puts the nursing care plan into action

42. Collaborative nursing interventions:

    A. are based on the written instructions of another professional.
    B. are determined solely by the nurse and patient.
    C. reflect the overlapping responsibilities of health care personnel.
    D. require supervision by the doctor.

43. Which nurse might need assistance when implementing the described nursing intervention?

A. Astrid, who has to insert a nasogastric tube in a patient; a procedure she has performed numerous times
B. Bill, who has to transfer a patient to a stretcher; the patient is in great pain
C. Mary, a member of the I.V. team, who has to start an I.V. on a patient requiring I.V. antibiotics
D. Sharon, who weighs 110 lb (50 kg), and has to position a patient who weighs 98 lb (45 kg)

44. When determining the teaching strategy that is best suited to the patient and the material to be learned, which intervention skill is the nurse using?

A. Cognitive
B. Effective
C. Interpersonal
D. Psychomotor

45. A group of nurses, doctors, and a health care administrator collaborate to detail the actions nurses are able to execute when a patient is undergoing I.V. heparin therapy. This example best describes:

A. a critical pathway.
B. a protocol.
C. a standardized nursing care plan.
D. an independent nursing action.

46. When assessing a patient's level of pain, which type of nursing intervention is the nurse performing?

A. Collaborative
B. Dependent
C. Independent
D. Professional

47. Which of the following is a true statement with regard to delegation of nursing interventions to unlicensed assistive personnel (UAP)?

A. A UAP determines which nursing interventions can be delegated.
B. Delegation of nursing interventions to a UAP resides with the doctor.
C. Rarely should a nurse delegate nursing interventions to a UAP.
D. The nurse is responsible and accountable for the outcome.

48. When a nurse and a respiratory therapist determine that 1 hour prior to each meal is the ideal time for a patient's respiratory therapy, which type of nursing intervention is the nurse performing?

    A. Codependent
    B. Collaborative
    C. Independent
    D. Professional

49. Dependent nursing interventions:

    A. are actions for which the nurse isn't accountable.
    B. are individualized for the patient by nurses.
    C. are primarily collegial in nature.
    D. need to be clarified by the nurse if questionable.

50. When reassuring a patient during an uncomfortable procedure, which intervention skill is the nurse using?

    A. Cognitive
    B. Effective
    C. Interpersonal
    D. Psychomotor

51. Which of the following behaviors by the nurse, Paula Smith, demonstrates that she understands the elements of effective charting?

    A. She documents giving a patient's medication after administering the medication.
    B. She documents the following about her patient: "appetite good this morning".
    C. She signs her charting as follows: P. Smith.
    D. She writes in the nurses' notes with a no. 2 pencil.

52. The type of patient record in which each member of the health care team or department makes notations about patient health problems that are integrated throughout the patient's chart is a:

    A. computer record.
    B. department record.
    C. problem-oriented medical record.
    D. source-oriented medical record.

53. What is a disadvantage of computerized documentation of the nursing process?

    A. Accuracy
    B. Concern for privacy
    C. Legibility
    D. Rapid communication

54. Which nursing documentation entry reflects a potential legal risk?

    A. "Patient's breath sounds are clear and equal bilaterally."
    B. "Patient's right antecubital space has needle tracks."
    C. "Patient states, 'I'm happy that I'm going home today.'"
    D. "Patient transferred from the bed to chair with one assist."

55. What should the nurse do after making a charting error in the nurses' notes?

    A. Draw a line through the error and write "Error" and her initials above it.
    B. Obliterate the mistake with a black felt pen.
    C. Recopy the page of nurses' notes and start over.
    D. Report the incident immediately to the head nurse.

56. During a change-of-shift report, it would be important for the nurse relinquishing responsibility for care of the patient to communicate which of the following facts to the nurse assuming responsibility for care of the patient?

    A. That the patient had a nasogastric tube removed 2 hours ago
    B. That the patient stated, "My pain is much better today than yesterday"
    C. That the patient's barium enema performed 3 days ago was negative
    D. That the patient's family came to visit this morning

57. A patient's participation would most likely be encouraged in:

    A. change-of-shift report.
    B. nursing care conference.
    C. nursing grand rounds.
    D. staff meeting.

58. Independent nursing interventions:

    A. are also known as standing orders.
    B. are initiated based on the nurse's own knowledge and skill.
    C. consist primarily of health education.
    D. relate directly to the patient's disease process.

59. Using the nursing process helps provide patient care that is:

    A. individualized.
    B. repetitious.
    C. standardized.
    D. unorthodox.

60. In the evaluation process, the nurse should:

    A. be certain that the diagnosis is amenable to medical care.
    B. have three other nurses confirm the accuracy of the diagnosis.
    C. look for patterns and deviations from normal.
    D. use only subjective data to infer problems.

61. The evaluative statement "At 0300, patient stated, 'It is easier to breathe,' rated pain a '3' on a scale of 1 to 10, and coughed effectively" is missing which of the following elements?

    A. Conclusion
    B. Decision
    C. Outcome
    D. Supporting data

62. The head nurse wants to determine how the size and location of the nursing station influences the delivery of nursing care on the unit. This is an example of which type of evaluation?

    A. Outcome
    B. Peer
    C. Process
    D. Structure

63. Which statement does *not* describe an appropriate guideline for writing a nursing diagnosis?

    A. State the diagnosis in terms of a problem, not a need.
    B. Use medical terminology to describe the probable cause of the patient's response.
    C. Use nursing terminology to describe the patient response.
    D. Use statements that assist in planning the independent nursing interventions.

64. The evaluation phase of the nursing process:

    A. changes continually.
    B. directs the expected patient outcomes.
    C. is a stable, systematic method of developing nursing care plans.
    D. isn't always necessary.

65. Which of the following statements about the term *nursing process* is true?

    A. It is used only in the United States.
    B. It originated with Florence Nightingale.
    C. It was first used by Lydia Hall in 1955.
    D. It was initiated by the National League for Nursing in 1983.

66. Cholelithiasis is a:

    A. collaborative problem.
    B. medical diagnosis.
    C. nursing diagnosis.
    D. potential problem.

SITUATION: *Gary Stellato is a 25-year-old male who has been vomiting and having diarrhea for 48 hours.*

    Questions 67 to 70 relate to this situation.

67. Which statement is the most appropriate goal for a nursing diagnosis of diarrhea?

    A. "The patient will experience decreased frequency of bowel evacuation."
    B. "The patient will provide a stool specimen for culture and sensitivity."
    C. "The patient will receive antidiarrheal medication."
    D. "The patient will save all stools for inspection by the nurse."

68. Which of the following is the most important purpose of planning care with this patient?

    A. Development of a standardized nursing care plan
    B. Expansion of the current taxonomy of nursing diagnosis
    C. Provision of individualized patient care
    D. Incorporation of both nursing and medical diagnoses in patient care

69. When selecting appropriate nursing interventions for Mr. Stellato, the nurse must remember that nursing interventions should be:

    A. achievable with resources available to the nurse and patient.
    B. carried out under the supervision of a doctor.
    C. chosen disregarding the patient's values and beliefs.
    D. oriented primarily toward tasks and mechanical procedures.

70. Where would the nurse most likely find Mr. Stellato's vital signs in a source-oriented medical record?

    A. Graphic sheet
    B. Diagnostic reports
    C. Nurses' notes
    D. Doctor's progress notes

ANSWER SHEET

	A B C D		A B C D		A B C D
1	○ ○ ○ ○	25	○ ○ ○ ○	49	○ ○ ○ ○
2	○ ○ ○ ○	26	○ ○ ○ ○	50	○ ○ ○ ○
3	○ ○ ○ ○	27	○ ○ ○ ○	51	○ ○ ○ ○
4	○ ○ ○ ○	28	○ ○ ○ ○	52	○ ○ ○ ○
5	○ ○ ○ ○	29	○ ○ ○ ○	53	○ ○ ○ ○
6	○ ○ ○ ○	30	○ ○ ○ ○	54	○ ○ ○ ○
7	○ ○ ○ ○	31	○ ○ ○ ○	55	○ ○ ○ ○
8	○ ○ ○ ○	32	○ ○ ○ ○	56	○ ○ ○ ○
9	○ ○ ○ ○	33	○ ○ ○ ○	57	○ ○ ○ ○
10	○ ○ ○ ○	34	○ ○ ○ ○	58	○ ○ ○ ○
11	○ ○ ○ ○	35	○ ○ ○ ○	59	○ ○ ○ ○
12	○ ○ ○ ○	36	○ ○ ○ ○	60	○ ○ ○ ○
13	○ ○ ○ ○	37	○ ○ ○ ○	61	○ ○ ○ ○
14	○ ○ ○ ○	38	○ ○ ○ ○	62	○ ○ ○ ○
15	○ ○ ○ ○	39	○ ○ ○ ○	63	○ ○ ○ ○
16	○ ○ ○ ○	40	○ ○ ○ ○	64	○ ○ ○ ○
17	○ ○ ○ ○	41	○ ○ ○ ○	65	○ ○ ○ ○
18	○ ○ ○ ○	42	○ ○ ○ ○	66	○ ○ ○ ○
19	○ ○ ○ ○	43	○ ○ ○ ○	67	○ ○ ○ ○
20	○ ○ ○ ○	44	○ ○ ○ ○	68	○ ○ ○ ○
21	○ ○ ○ ○	45	○ ○ ○ ○	69	○ ○ ○ ○
22	○ ○ ○ ○	46	○ ○ ○ ○	70	○ ○ ○ ○
23	○ ○ ○ ○	47	○ ○ ○ ○		
24	○ ○ ○ ○	48	○ ○ ○ ○		

ANSWERS AND RATIONALES

1. *Correct answer:* **B**

The nursing process is goal-oriented. It is also systematic, patient-centered, and dynamic.

Taxonomic level: Knowledge

2. *Correct answer:* **B**

The order of the nursing process is assessing, diagnosing, planning, implementing, and evaluating.

Taxonomic level: Knowledge

3. *Correct answer:* **D**

The trial-and-error method of problem solving isn't systematic (as is the scientific method of problem solving), routine, or based on inner prompting (as is the intuitive method of problem solving).

Taxonomic level: Knowledge

4. *Correct answer:* **A**

Deductive reasoning moves from general principles to the collection of specific data or information that confirm or negate a hypothesis. Inductive reasoning involves forming generalizations from a set of facts or observations. Insightful and rational methods are not used to confirm or negate a hypothesis.

Taxonomic level: Knowledge

5. *Correct answer:* **A**

The outcome, or "product," of the planning phase of the nursing process is a nursing care plan.

Taxonomic level: Knowledge

6. *Correct answer:* **A**

The nursing process can be used with patients of any age, at any point on the wellness-illness continuum, in a variety of settings, and across specialty areas. Therefore, its range of applications is broad.

Taxonomic level: Knowledge

7. *Correct answer:* **C**

Objective data are also known as signs, or overt data.

Taxonomic level: Knowledge

8. *Correct answer:* **A**

The nurse uses data or information obtained from the assessment phase of the nursing process primarily to ascertain the patient's responses to health problems.

Taxonomic level: Knowledge

9. *Correct answer:* **B**

The patient is the primary source of data collection in the assessment phase of the nursing process; the other options are secondary sources.

Taxonomic level: Knowledge

10. *Correct answer:* **D**

Subjective data are apparent only to the person affected and can be described or verified by only that person. Therefore, only the patient can describe or verify whether he's nauseous.

Taxonomic level: Knowledge

11. *Correct answer:* **C**

Focused assessments gather data or information about specific patient health problems identified during comprehensive assessments. Although focused assessments can be included during comprehensive assessments, they routinely are performed as part of ongoing data collection.

Taxonomic level: Knowledge

12. *Correct answer:* **B**

The primary source, or the patient, is Jason; his mother is a secondary source. The data (Jason vomited 8 ounces of his formula this morning) is objective because it can be perceived by the senses, verified by another person observing the same patient, and tested against accepted standards or norms.

Taxonomic level: Comprehension

13. *Correct answer:* **A**

Objective data (such as +4 patellar reflexes in both of the patient's legs) are data that can be perceived by the senses, verified by another person observing the same patient, and tested against accepted standards or norms. Subjective data (for example, tinnitus, numbness, and vertigo) are apparent only to the person affected and can be described and verified by only that person.

Taxonomic level: Knowledge

14. *Correct answer:* **B**

Potentially life-threatening problems, such as diminished breath sounds, are high priority because they pose the greatest threat to the patient's well-being. Thus, they require immediate communication to another member of the health care team.

Taxonomic level: Application

15. *Correct answer:* **C**

The nursing diagnosis of impaired gas exchange is listed in the NANDA taxonomy under the human response pattern of exchanging, which involves mutual giving and receiving.

Taxonomic level: Knowledge

16. *Correct answer:* **C**

The etiology provides one or more probable causes of the patient's health problem. As such, it also gives direction to the required nursing interventions necessary to treat the problem. Unless the risk factors are correctly identified, the nursing interventions selected to treat the patient's health problem may be inefficient, ineffective, or both.

Taxonomic level: Comprehension

17. *Correct answer:* **D**

A data cluster is a grouping of significant patient data or information that points to the existence of a particular patient health problem.

Taxonomic level: Comprehension

18. *Correct answer:* **D**

The patient is exhibiting signs and symptoms (defining characteristics) and the etiology (risk factors) of an actual nursing diagnosis; in this instance, the actual nursing diagnosis is fatigue.

Taxonomic level: Comprehension

19. *Correct answer:* **B**

An inference is a nurse's judgment or interpretation of cues (any piece of data or information that influences a nurse's decisions). In this instance, the nurse infers from the patient's heart rate of 60 beats/minute that the patient has bradycardia.

Taxonomic level: Application

20. *Correct answer:* **C**

According to Maslow's hierarchy of basic human needs, physiologic needs (such as an effective breathing pattern) must be met before lower needs (such as safety, love and belonging, self-esteem, and self-actualization) can be met. Therefore, physiologic needs have the highest priority.

Taxonomic level: Application

21. *Correct answer:* **B**

In the diagnosing phase of the nursing process, the nurse compares data against standards or norms, which are generally accepted rules, models, patterns, or measures that can be used for comparing data in the same class or category.

Taxonomic level: Knowledge

22. *Correct answer:* **B**

Although diagnosing is basically the nurse's responsibility, input from the patient is essential to formulate the correct nursing diagnosis.

Taxonomic level: Knowledge

23. *Correct answer:* **A**

Only anxiety is a NANDA-approved nursing diagnosis. Diabetes, heart failure, and myocardial infarction are medical diagnoses.

Taxonomic level: Comprehension

24. *Correct answer:* **D**

Premature closure is making a judgment based on only one or two pieces of data or information.

Taxonomic level: Comprehension

25. *Correct answer:* **C**

Measuring and recording the patient's total fluid intake and output every shift will determine whether the goal or expected outcome of "total fluid intake equals total fluid output" has been met.

Taxonomic level: Application

26. *Correct answer:* **D**

Standardized nursing care plans are preplanned, preprinted guides for the care of groups of patients with common needs. However, unless a nurse uses the checklists, blank lines, or empty spaces located within a standardized nursing care plan to individualize it, the unique needs of the patient may not be addressed.

Taxonomic level: Knowledge

27. *Correct answer:* **C**

Expected outcomes are specific, measurable criteria used to evaluate goal achievement. The phrases "an adequate amount," "less edematous [swollen]," and "sleep well" are vague and not measurable.

Taxonomic level: Application

28. *Correct answer:* **D**

In the planning phase of the nursing process, the nurse works with the patient and family to formulate patient goals or expected outcomes and to design the nursing interventions necessary to achieve those goals.

Taxonomic level: Knowledge

29. *Correct answer:* **C**

Discharge planning should commence upon admission because the average length of stay for patients in acute-care hospitals is becoming increasingly short.

Taxonomic level: Knowledge

30. *Correct answer:* **A**

Short-term goals are best for patients who require health care for a short time (usually less than a week, such as for an appendectomy). Long-term goals are often used for patients who require health care for a longer period of time (usually more than a week, such as for burns, a spinal cord injury, and heart failure).

Taxonomic level: Application

31. *Correct answer:* **B**

The scientific reason for selecting a specific nursing intervention supported by clinical research is called a rationale. A criterion is a method of judging something. A strategy is a plan for approaching a problem or task. A theory is a hypothesis that isn't based on actual knowledge.

Taxonomic level: Knowledge

32. *Correct answer:* **C**

The evaluative statement—"Goal met"— is consistent with the supporting data— "Patient was able to transfer from the bed to a wheelchair without assistance"— and demonstrates goal achievement.

Taxonomic level: Analysis

33. *Correct answer:* **A**

Outcome evaluation focuses on measurable changes in the health status of the patient or the end results of nursing care.

Taxonomic level: Knowledge

34. *Correct answer:* **D**

In the evaluating phase of the nursing process, the nurse and patient together measure how well the patient has achieved the goals and expected outcomes in the nursing care plan. In the analysis phase, the nurse formulates a nursing diagnosis; in the implementing phase, the nurse carries out the nursing interventions; and in the planning phase, the nurse identifies available resources.

Taxonomic level: Knowledge

35. *Correct answer:* **D**

The conclusion, "Goal partially met," is consistent with the supporting data, and "patient able to state only three warning signals of cancer" demonstrates partial goal achievement.

Taxonomic level: Analysis

36. *Correct answer:* **D**

The nurse should use the method indicated by the expected outcome as a guide to collect data to ensure that correct conclusions can be drawn about goal achievement.

Taxonomic level: Application

37. *Correct answer:* **C**

Retrospective means relating to a past event. In this instance, information obtained from the questionnaire and used to ascertain the patient's level of satisfaction with the nursing care received in the agency was obtained after (retrospectively of) the patient's recent hospitalization.

Taxonomic level: Knowledge

38. *Correct answer:* **C**

If a goal isn't met, the nurse should review part or all of the entire nursing care plan and critique the steps of the process involved in the development of the goal.

Taxonomic level: Comprehension

39. *Correct answer:* **C**

Process evaluation focuses on the nature and sequence of activities used by nurses in implementing the nursing care plan.

Taxonomic level: Knowledge

40. *Correct answer:* **B**

Quality improvement is the commitment and approach to continuously improve every process in every part of an organization with the intent of meeting or exceeding customer expectations.

Taxonomic level: Comprehension

41. *Correct answer:* **D**

In the implementation phase of the nursing process, the nurse carries out all nursing interventions developed during the planning phase. In the assessment phase, the nurse determines the patient's health status; in the planning phase, she identifies available resources; in the evaluation phase, she measures goal achievement.

Taxonomic level: Knowledge

42. *Correct answer:* **C**

Collaborative nursing interventions involve nursing interventions performed jointly by nurses and other members of the health care team.

Taxonomic level: Knowledge

43. *Correct answer:* **B**

A nurse may determine the need for assistance when implementing a nursing intervention with assistance would reduce stress on the patient.

Taxonomic level: Application

44. *Correct answer:* **A**

Cognitive nursing intervention skills include problem solving, decision making, critical thinking, and creative thinking.

Taxonomic level: Knowledge

45. *Correct answer:* **B**

Protocols are preprinted, preplanned, detailed guides of nursing interventions that nurses are able to execute in specific situations for a particular group of patients.

Taxonomic level: Knowledge

46. *Correct answer:* **C**

Independent nursing interventions involve actions that nurses initiate based on their own knowledge and skill and without the direction or supervision of another member of the health care team.

Taxonomic level: Comprehension

47. *Correct answer:* **D**

Delegation is the transfer of responsibility for the performance of an activity from one individual to another while retaining accountability for the outcome.

Taxonomic level: Comprehension

48. *Correct answer:* **B**

Collaborative nursing interventions involve actions carried out with another member of the health care team.

Taxonomic level: Knowledge

49. *Correct answer:* **D**

Nurses are accountable for the dependent nursing interventions they implement. As such, they are responsible for clarifying any questionable dependent nursing interventions with the doctor.

Taxonomic level: Knowledge

50. *Correct answer:* **C**

Interpersonal nursing intervention skills involve the ability to communicate such things as caring, comfort, and support to the patient.

Taxonomic level: Comprehension

51. *Correct answer:* **A**

A nurse should record a nursing intervention (for example, administering a medication) after performing the nursing intervention (not before). Recording should also be done in black ink (not pencil), be complete, and be signed with the nurse's full name and title.

Taxonomic level: Comprehension

52. *Correct answer:* **C**

In problem-oriented medical records, data or information about patient health problems — whether gathered by doctors, nurses, or other health care team members involved in the patient's care — is integrated throughout the chart, rather than organized according to the source of the data or information.

Taxonomic level: Knowledge

53. *Correct answer:* **B**

A patient's privacy may be violated if security measures, such as passwords and firewalls, aren't used properly or if policies and procedures aren't in place that determine what type of information can be retrieved, by whom, and for what purpose.

Taxonomic level: Knowledge

54. *Correct answer:* **B**

If not correct, the inference that the patient has needle tracks and has abused or is currently abusing drugs places the nurse at risk for libel (defamation by means of print, writing, or pictures).

Taxonomic level: Application

55. *Correct answer:* **A**

Recording errors shouldn't be changed in any way that raises doubt about the nursing care given or the charting error made (such as erasing or blotting them out with correction fluid).

Taxonomic level: Knowledge

56. *Correct answer:* **A**

The change-of-shift report should include significant recent changes in the patient's condition that the nurse assuming responsibility for care of the patient will need to monitor. The other options aren't critical enough to include in the report.

Taxonomic level: Application

57. *Correct answer:* **C**

In the nursing grand rounds format, a group of nurses visit selected patients at bedside, discuss each patient's care, and encourage each patient to participate in the discussion.

Taxonomic level: Knowledge

58. *Correct answer:* **B**

Independent nursing interventions involve actions nurses initiate based on their own knowledge and skill and without the direction or supervision of another member of the health care team.

Taxonomic level: Knowledge

59. *Correct answer:* **A**

The nursing process results in the development of a nursing care plan that reflects the unique, individualized needs of the patient.

Taxonomic level: Knowledge

60. *Correct answer:* **C**

To formulate a correct nursing diagnosis with the patient during the diagnostic phase of the nursing process, the nurse must look for patterns and deviations from normal with which to form a data cluster.

Taxonomic level: Knowledge

61. *Correct answer:* **A**

An evaluative statement includes two parts: the conclusion (either "Goal met," "Goal partially met," or "Goal not met") and supporting data for the conclusion.

Taxonomic level: Knowledge

62. *Correct answer:* **D**

Structure evaluation focuses on the environment in which nursing care is provided.

Taxonomic level: Knowledge

63. *Correct answer:* **B**

The nursing diagnosis is a statement about a patient's actual or potential health problem that is within the scope of independent nursing intervention.

Taxonomic level: Knowledge

64. *Correct answer:* **A**

By continually evaluating whether the expected patient outcomes have been accomplished, the nurse can change the nursing care plan to reflect changes in the patient's condition; thus, evaluation results in a constantly changing nursing process. Evaluation is always a phase in the nursing process. It helps the nurse measure (not direct) expected patient outcomes.

Taxonomic level: Knowledge

65. *Correct answer:* **C**

Lydia Hall first used the term *nursing process* in 1955.

Taxonomic level: Knowledge

66. *Correct answer:* **B**

Cholelithiasis is a medical diagnosis and not a health problem that nurses are licensed to treat or that can be resolved by independent nursing interventions.

Taxonomic level: Knowledge

67. *Correct answer:* **A**

The goal is the opposite, healthy response of the problem statement (diagnostic label) of the nursing diagnosis; in this instance, the problem statement is diarrhea.

Taxonomic level: Application

68. *Correct answer:* **C**

To be effective, the nursing care plan developed in the planning phase of the nursing process must reflect the individualized needs of Mr. Stellato.

Taxonomic level: Knowledge

69. *Correct answer:* **A**

If a nursing intervention requires the use of resources unavailable to the nurse and patient, it may not be effective in treating the patient's health problem.

Taxonomic level: Knowledge

70. *Correct answer:* **A**

The graphic sheet is a specialized flow sheet in which patient variables, such as temperature, pulse, respiration, and blood pressure, are located in a manner providing a quick reflection of the patient's condition.

Taxonomic level: Knowledge

FOUNDATIONS FOR NURSING PRACTICE

QUESTIONS

1. The person who is considered the founder of modern nursing and who established the theoretical base for nursing is:

    A. Ann Preston.
    B. Florence Nightingale.
    C. Jean Watson.
    D. Linda Richards.

2. A key element of Florence Nightingale's nursing practice was:

    A. maintaining a maximal level of wellness.
    B. promoting the patient's self-care needs.
    C. recognizing nutrition as an important part of nursing care.
    D. addressing the patient's adaptation to health problems.

3. The first trained nurse in the United States to start the practice of keeping records and writing orders was:

    A. Isabel Hampton Robb.
    B. Lavinia Dock.
    C. Linda Richards.
    D. Louisa Lee Schuyler.

4. Which of the following statements accurately describes a nursing school before 1900 in the United States?

    A. Faculty were highly educated, with most having master's degrees.
    B. Most classes were taught by doctors.
    C. Most nursing schools were in colleges or universities.
    D. Student nurses worked only 10 hours per week in clinics.

5. In the 1890s, the Henry Street Settlement House was established in New York City. Its purpose was to provide:

    A. education for the poor and a place for social gatherings.
    B. emergency care for war victims.
    C. continuing education for doctors and nurses.
    D. nursing education for students.

6. The American Red Cross was founded by:

    A. Clara Barton.
    B. Florence Nightingale.
    C. Lavinia Dock.
    D. Linda Richards.

7. The crusade to stop the inhumane treatment of patients in mental hospitals was lead by:

    A. Benjamin Franklin.
    B. Clara Barton.
    C. Dorothea Dix.
    D. Louisa Schuyler.

8. Prior to 1950, the majority of nursing education in the United States took place in:

    A. colleges and universities.
    B. community colleges.
    C. correspondence courses.
    D. hospital-based diploma programs.

9. There are three primary types of nursing programs (Associate Degree in Nursing, diploma, Bachelor of Science in Nursing) leading to registered nurse status in the United States. Graduates must complete which of the following processes?

    A. Certification
    B. Licensure
    C. Reciprocity
    D. Verification

10. In the United States, a professional organization whose originating organization was founded in the late 1800s and whose members are state nursing associations, with individual nurses belonging to the state organization, is called the:

    A. American Association of Colleges of Nursing (AACN).
    B. American Nurses Association (ANA).
    C. Canadian Nurses Association (CNA).
    D. National League for Nursing (NLN).

11. The American Nurses Association *Standards of Clinical Practice* was written in 1991 to define the activities of nurses that are specific and unique to nursing. These standards laid the foundation for practice in professional nursing in all settings. The standards are based on:

    A deductive reasoning.
    B. dependent nursing.
    C. nursing process.
    D. simple to complex activities.

12. The nurse has organized an immunization clinic for healthy babies and preschool children. This would be an example of what level of preventative health care?

    A. Curative
    B. Primary
    C. Secondary
    D. Tertiary

13. The nurse is caring for a patient in the hospital after the surgical removal of his gallbladder. This would be an example of what level of preventative health care?

    A. Curative
    B. Primary
    C. Secondary
    D. Tertiary

14. Which of the following statements indicates a trend affecting professional nursing practice today?

    A. Hospitals are changing the direct patient contact mix by hiring more nonprofessional personnel, thus decreasing the number of professional nurses.
    B. Nursing is moving back to internships instead of formal education.
    C. Psychologists are conducting nursing research instead of using nurses.
    D. Very few nurses work in the community; almost all work in hospitals.

15. A nurse is a lifelong learner. The most important reason for this is:

    A. accrediting agencies, such as the National League for Nursing, require it.
    B. employers require it to stay employed.
    C. doctors need to be ensured the nurse is competent in nursing.
    D. nurses must remain current in nursing research, skills, and knowledge.

16. Which of the following nursing positions requires an advanced degree?

    A. Critical care nurse
    B. Oncology nurse
    C. Nurse practitioner
    D. Public health nurse

17. Maslow's hierarchy of basic needs is a common theory that nursing education incorporates to explain the basic needs of people. What need must be met before the person can focus on safety and security?

    A. Love and belonging
    B. Physiological
    C. Self-actualization
    D. Self-esteem

18. Health care delivery in the future must consider what major trend?

    A. Decreasing competition among hospitals
    B. Decreasing emphasis on technology
    C. Increasing number of nurses in the profession
    D. Increasing population over age 70

19. The nurse performs many roles in the practice of nursing. Which role is defined as "the protection of human or legal rights and the securing of quality care for each patient"?

    A. Advocator
    B. Communicator
    C. Counselor
    D. Leader

20. Critical thinking is essential to a nurse because it:

    A. is constantly used for solving problems and making sound judgments.
    B. becomes vital in developing nursing knowledge.
    C. forms the basis for the professional code of ethics of nursing.
    D. is needed to successfully pass the licensure examination.

21. Academic preparation for the licensed practical nurse (LPN) degree is usually:

    A. one year of technical school.
    B. a two-year community college degree.
    C. a three-year hospital program.
    D. a four-year university degree.

22. The nurse is teaching a diabetic patient how to inject insulin and the dosages necessary for optimal control. This would be an example of what level of health care?

    A. Curative
    B. Primary
    C. Secondary
    D. Tertiary

23. A patient comes to the clinic with a diagnosis of high blood pressure, and the doctor asks the nurse to speak to the patient about his diet, especially cutting down on salt. The nurse's actions are an example of what level of health care?

    A. Primary
    B. Restorative
    C. Secondary
    D. Tertiary

24. A major purpose for the existence of professional nursing organizations is to:

    A. help set standards for the practice of nursing.
    B. keep data on members.
    C. provide a social outlet for members.
    D. regulate members' job responsibilities.

25. All registered nurses may function in all of the following capacities, *except*:

    A. participate in gathering data for nursing research.
    B. provide health teaching at time of discharge from the hospital.
    C. work in the operating room assisting the doctor.
    D. write prescriptions for common ailments.

26. Licensure to practice nursing is controlled by the:

    A. American Association of Colleges of Nursing (AACN).
    B. American Nurses Association (ANA).
    C. National League for Nursing (NLN).
    D. State Board of Nursing.

27. A master of science in nursing degree (MSN):

    A. is considered the level of entry education by the American Nurses Association.
    B. is granted upon successful completion of an Associate Degree in Nursing.
    C. may be acquired by completing continuing education programs.
    D. prepares the individual to function in expanded roles.

28. The model of care that assigns tasks instead of patients to caregivers is called:

    A. case management nursing.
    B. functional nursing.
    C. team nursing.
    D. total patient care nursing.

29. The United States federal program that provides funding for health care for people age 65 and older is:

    A. HMO.
    B. Medicaid.
    C. Medicare.
    D. Social Security.

30. The payment category that Medicare uses to fix fees for payment is called:

    A. CHAMPUS.
    B. diagnostic related group (DRG).
    C. HMO.
    D. preferred provider organization (PPO).

31. A risk in obtaining informed consent before surgery is:

    A. completing the informed consent papers before all preoperative test results are in the patient's chart.
    B. completing the informed consent papers 1 hour after the preoperative medication is given.
    C. completing the informed consent papers the evening before surgery.
    D. completing the informed consent papers with a family member present.

32. A diagnostic test is ordered that requires a signed informed consent. The nurse becomes very busy and she checks on the patient, discovering that the patient has already left for the test. The nurse goes to find the patient just as the procedure is completed. A procedure without the proper informed consent may result in what type of crime?

    A. Battery
    B. Challenge for cause
    C. Comparative negligence
    D. Slander

33. A nurse performs cardiopulmonary resuscitation in an emergency situation until the ambulance arrives to take the victim to the hospital. The nurse may be protected by what law in this case?

    A. Good Samaritan
    B. Minimum Standards
    C. Nurse Practice Act
    D. OSHA Act

34. After delivering a baby with special needs, the new mother tells the nurse that the man that assumes he's the father isn't and asks the nurse to tell the doctor. This medical dilemma is an example of:

    A. confidentiality.
    B. conflict of interest.
    C. negligence.
    D. slander.

35. A nurse caring for a two-year-old patient leaves the bedside without raising the side rails and the patient falls to the floor. What type of legal action could be used in a lawsuit?

    A. Battery
    B. Malpractice
    C. Negligence
    D. Slander

36. A law voted on by a legislative body of government is called:

    A. administrative law.
    B. common law.
    C. constitutional law.
    D. statutory law.

37. When a judge interprets a law and issues a decision, this is called:

    A. administrative law.
    B. common law.
    C. constitutional law.
    D. statutory law.

38. A nursing theory is a group of concepts that form a pattern of reality. A concept can be defined as:

    A. an idea.
    B. an outcome.
    C. a philosophy.
    D. a process.

39. A concept's most important purpose in professional nursing is to:

    A. determine nursing action that is appropriate in a variety of situations.
    B. help mentally store knowledge so it doesn't have to be written down.
    C. help recall facts in the clinical area.
    D. use Neuman's model in the work setting.

40. The four concepts common in all nursing theories are person, environment, health, and nursing. Of these four, the most important concept is:

    A. environment.
    B. health.
    C. nursing.
    D. person.

41. The theorist who believes that adaptation and manipulation of stressors
are needed to foster change is:

    A. Betty Neuman.
    B. Dorothea Orem.
    C. Martha Rogers.
    D. Sister Callista Roy.

42. The nursing model based on the belief that the individual has a need for
self-care actions that nursing can assist in meeting is called:

    A. adaptation.
    B. behavioral system.
    C. self-care deficit.
    D. transcultural.

43. A theory that can be traced to the 1930s and 1940s in which it is believed that
social, not biological, forces are the primary source of human behavior is:

    A. adaptation theory.
    B. general systems theory.
    C. role theory.
    D. self-care deficit theory.

44. The theorist whose theory can be defined as the development of a science
of humankind, incorporating the concepts of energy fields, openness, pattern, and organization, is:

    A. Dorothy Johnson.
    B. Hildegard Peplau.
    C. Martha Rogers.
    D. Myra Levine.

45. Betty Neuman's theory can be defined as:

    A. conceptualized nursing as a total-person approach.
    B. goal attainment through nurse-patient interactions.
    C. philosophy and science of caring and humanistic nursing.
    D. the nurse-patient relationship using a "deliberative" nursing ap-
proach.

46. A theorist whose major theme is the idea of transcultural nursing and caring nursing is:

    A. Dorothea Orem.
    B. Madeleine Leininger.
    C. Sister Callista Roy.
    D. Virginia Henderson.

47. A patient has been hospitalized after a car accident. He's planning to sue the driver of the vehicle that hit him. In this lawsuit, the patient is the:

    A. agitator.
    B. defendant.
    C. litigator.
    D. plaintiff.

48. The nurse makes two medication errors because she didn't use the five rights of medication administration and one patient has a reaction. This is an example of:

    A. assault.
    B. fraud.
    C. libel.
    D. malpractice.

49. The most common reason for a nursing license to be revoked is:

    A. developing a physical deficit.
    B. drug and alcohol abuse.
    C. not paying income taxes.
    D. writing bad checks.

50. An appropriate reason for the loss of a nursing license is:

    A. bankruptcy.
    B. decline in health.
    C. theft.
    D. therapeutic abortion.

51. The doctor orders a very large dose of narcotics for a terminally ill patient who hasn't previously been taking narcotics. The nurse knows that this dose could be fatal if administered. Legally, the nurse:

    A. can give the medication after she gets approval from the nursing supervisor.
    B. is responsible for any medication she administers.
    C. should give the medications, because the doctor is responsible for his order.
    D. should notify authorities immediately.

52. The nurse has made assignments that included a nurse assistant who is an unlicensed caregiver. Is the nurse legally responsible for the care the nurse assistant gives?

    A. Always because the nurse is her supervisor

    B. Never because the hospital employs both the nurse and the nurse assistant, so they are both responsible

    C. Not usually if the nurse assistant has been assigned tasks within the job description of a nurse assistant

    D. Possibly because the law is so different in each state that there is no clear answer

53. Nurses should be aware of the political process because:

    A. doctors will take over nursing therapies.

    B. managed care is becoming the leader in the pay process for care.

    C. nurses may be called to war if needed.

    D. some nursing practices are controlled by governmental decisions.

54. One source that is likely to be less biased when reporting political issues and statements would be the:

    A. *Federal Register.*

    B. local newspaper.

    C. senators' newsletter.

    D. weekly news magazine.

55. Sometimes a doctor or nurse may unintentionally harm a patient. This is called:

    A. libel.

    B. malpractice.

    C. negligence.

    D. nonmaleficence.

56. Fraud can be defined as:

    A. false accusations — written, printed, or typed — that are made with malicious intent.

    B. any words spoken with malice that are untrue and prejudicial to another person.

    C. the act of intentionally misleading or deceiving another person by any means.

    D. the omission of an act that a prudent person would have performed.

57. A nurse alters a hospital record because she knows an error has been committed. Her action may be classified as:

    A. fraud.
    B. libel.
    C. malpractice.
    D. slander.

58. A nurse gives legal testimony that is taken and recorded outside the courtroom. This is called a:

    A. certified testimony.
    B. contract.
    C. deposition.
    D. tort.

59. Law is defined as a standard of conduct established and enforced by the government. There are several types of law, and the practice of nursing is governed by:

    A. administrative law.
    B. civil law.
    C. common law.
    D. public law.

60. The Patient's Bill of Rights was published by the:

    A. American Hospital Association.
    B. American Nurses Association.
    C. Joint Committee of Hospitals.
    D. Occupational, Safety and Health Act.

61. Mr. Marsh wants his son designated to make his health decisions. What documentation is needed?

    A. A doctor's order
    B. A durable power of attorney for health care
    C. A living will signed by both parties
    D. A signed advance directive

ANSWER SHEET

	A B C D		A B C D		A B C D
1	○ ○ ○ ○	22	○ ○ ○ ○	43	○ ○ ○ ○
2	○ ○ ○ ○	23	○ ○ ○ ○	44	○ ○ ○ ○
3	○ ○ ○ ○	24	○ ○ ○ ○	45	○ ○ ○ ○
4	○ ○ ○ ○	25	○ ○ ○ ○	46	○ ○ ○ ○
5	○ ○ ○ ○	26	○ ○ ○ ○	47	○ ○ ○ ○
6	○ ○ ○ ○	27	○ ○ ○ ○	48	○ ○ ○ ○
7	○ ○ ○ ○	28	○ ○ ○ ○	49	○ ○ ○ ○
8	○ ○ ○ ○	29	○ ○ ○ ○	50	○ ○ ○ ○
9	○ ○ ○ ○	30	○ ○ ○ ○	51	○ ○ ○ ○
10	○ ○ ○ ○	31	○ ○ ○ ○	52	○ ○ ○ ○
11	○ ○ ○ ○	32	○ ○ ○ ○	53	○ ○ ○ ○
12	○ ○ ○ ○	33	○ ○ ○ ○	54	○ ○ ○ ○
13	○ ○ ○ ○	34	○ ○ ○ ○	55	○ ○ ○ ○
14	○ ○ ○ ○	35	○ ○ ○ ○	56	○ ○ ○ ○
15	○ ○ ○ ○	36	○ ○ ○ ○	57	○ ○ ○ ○
16	○ ○ ○ ○	37	○ ○ ○ ○	58	○ ○ ○ ○
17	○ ○ ○ ○	38	○ ○ ○ ○	59	○ ○ ○ ○
18	○ ○ ○ ○	39	○ ○ ○ ○	60	○ ○ ○ ○
19	○ ○ ○ ○	40	○ ○ ○ ○	61	○ ○ ○ ○
20	○ ○ ○ ○	41	○ ○ ○ ○		
21	○ ○ ○ ○	42	○ ○ ○ ○		

ANSWERS AND RATIONALES

1. *Correct answer:* **B**

Born in 1820, Florence Nightingale is considered the person who established nursing education in England and changed the image of nursing to that of a respectable calling. Dr. Ann Preston started the Women's Hospital in 1861. Jean Watson is a modern theorist. Linda Richards is considered the first American-trained nurse, having graduated in 1872 from the New England Hospital for Women and Children.

Taxonomic level: Knowledge

2. *Correct answer:* **C**

Nightingale established a training school in England and wrote books about health care. She recognized nutrition and occupational and recreational therapy as important and established standards for hospital management. Promoting self-care, maintaining a maximum level of wellness, and addressing adaptation weren't part of Nightingale's nursing practice.

Taxonomic level: Knowledge

3. *Correct answer:* **C**

Linda Richards was the first trained nurse in the United States to start the practice of keeping records and writing orders. She graduated from the New England Hospital for Women and Children in 1872. Louisa Lee Schuyler, a social reformer, opened Bellevue Hospital, one of the first Nightingale-inspired nurse training schools in America. Lavinia Dock, a nursing leader and suffragist, supported founding the Nurses Associated Alumnae of Training Schools, which reorganized into the American Nurses Association. Isabel Hampton Robb was one of the founders of the National League for Nursing and an 1893 graduate of Bellevue Hospital.

Taxonomic level: Knowledge

4. *Correct answer:* **B**

Before 1900, nurses were educated in an apprenticeship method in hospitals. Formal classes were rare, and most were taught by doctors. Students usually had half a day off to go to church. Faculty were usually the ward nurses under whom the students worked, and they had no more formal education than what the students themselves were receiving.

Taxonomic level: Knowledge

5. *Correct answer:* **A**

Lillian Wald founded the Henry Street Settlement House in New York City in 1893. It provided education for the poor and a place for social gatherings. The purpose of the Settlement House didn't include emergency care, continuing education for doctors and nurses, or nursing education for students.

Taxonomic level: Knowledge

6. *Correct answer:* **A**

Clara Barton founded the American Red Cross in 1882. Florence Nightingale is considered the founder of the modern nursing movement. Lavinia Dock started the Visiting Nurses Service in 1886. Linda Richards is one of the first American-trained nurses.

Taxonomic level: Knowledge

7. *Correct answer:* **C**

Dorothea Dix was a humanitarian who worked for improving prison conditions and care of the mentally ill. She helped establish 30 mental institutions in the United States and Canada. Clara Barton founded the American Red Cross. Benjamin Franklin was an early inventor and politician. Louisa Schuyler opened Bellevue Hospital.

Taxonomic level: Knowledge

8. *Correct answer:* **D**

Most nursing education before 1950 was at the diploma level. At one time, there were more than 1,000 hospital-based diploma programs for nursing. Community colleges began giving the Associate Degree in Nursing in the 1950s. College and university nursing education has increased since 1950.

Taxonomic level: Knowledge

9. *Correct answer:* **B**

Whatever nursing program is completed, all nursing graduates take the same licensing exam to become registered nurses. Reciprocity is when a nurse asks for licensure in another state after successfully obtaining licensure in one state. Verification and certification aren't needed for licensure.

Taxonomic level: Knowledge

10. *Correct answer:* **B**

The ANA is a professional nursing organization whose members are state nursing organizations. The NLN, established in 1952, has non-nurse members. The

AACN is an organization for colleges with Bachelor of Science in Nursing and higher-level nursing programs. The CNA is the Canadian nursing organization.

Taxonomic level: Knowledge

11. *Correct answer:* **C**

The standards are based on the nursing process of assessment, diagnosis, outcome identification, planning, implementation, and evaluation. The other answers don't pertain to the standards.

Taxonomic level: Knowledge

12. *Correct answer:* **B**

The primary level of preventative health care focuses on health promotion. Secondary health care focuses on health maintenance; tertiary health care focuses on rehabilitation. There is no curative level of preventative health care.

Taxonomic level: Comprehension

13. *Correct answer:* **C**

The secondary level of preventative health care focuses on health maintenance. Primary health care focuses on health promotion; tertiary health care focuses on rehabilitation. There is no curative level of preventative health care.

Taxonomic level: Comprehension

14. *Correct answer:* **A**

Hospitals are changing the direct patient contact mix by adding more nonprofessional personnel. Options B, C, and D are just the opposite of today's trends because nursing is moving increasingly to formal education, there are more nurse researchers actively working, and community nursing continues to grow.

Taxonomic level: Application

15. *Correct answer:* **D**

Nursing knowledge is ever changing; thus, nurses need to read and attend seminars and conferences to stay current in the profession. Accrediting agencies and doctors don't require nurses to continue learning. There are specific nursing positions and certain states that require a certain number of hours of continuing education to stay employed or to renew an RN license.

Taxonomic level: Comprehension

16. *Correct answer:* **C**

To use the title of nurse practitioner, an advanced degree is required. Nurses with either a diploma, an Associate Degree in Nursing, or a Bachelor of Science in Nursing may work in public health, critical care, or oncology. There may be short courses needed for specialties.

Taxonomic level: Knowledge

17. *Correct answer:* **B**

Maslow's hierarchy of basic needs provides a framework for nursing assessment and the relationships of basic needs and priorities of care. The needs listed from simple to complex are physiological, safety and security, love and belonging, self-esteem, and self-actualization.

Taxonomic level: Knowledge

18. *Correct answer:* **D**

People are living longer; thus, health care delivery will see more multiple chronic illnesses in people over age 70. There will probably be an increased emphasis on technology and competition among hospitals. A shortage of nurses (not an over-supply) is being predicted.

Taxonomic level: Knowledge

19. *Correct answer:* **A**

The advocator role provides for "the protection of human or legal rights and the securing of quality care for each patient." The leader role is defined as assertive, self-confident practice in nursing when providing or supervising care. The communicator role is the use of effective interpersonal and therapeutic communication skills to facilitate health care. The counselor role is the use of therapeutic communication to provide information to make referrals and facilitate problem solving.

Taxonomic level: Comprehension

20. *Correct answer:* **A**

Nurses need critical thinking every day to help solve problems and make sound judgments and good decisions for their patients. This is vital in caring for patients in any setting. The other answers are only partially correct.

Taxonomic level: Comprehension

21. *Correct answer:* **A**

A student will attend a practical nurse program for one year and then become eligible to take the LPN licensure examination. Students completing programs that are two, three, or four years in length may take the licensing examination to become a registered nurse.

Taxonomic level: Knowledge

22. *Correct answer:* **D**

The tertiary level of care focuses on the rehabilitation of the patient to a maximum level of functioning. The primary level of care focuses on health promotion; the secondary level focuses on health maintenance. Curative isn't a level of health care.

Taxonomic level: Knowledge

23. *Correct answer:* **D**

The tertiary level of care focuses on the rehabilitation of the patient to a maximum level of functioning. The primary level of care focuses on health promotion; the secondary level focuses on health maintenance. Restorative isn't a level of health care.

Taxonomic level: Comprehension

24. *Correct answer:* **A**

Nursing organizations such as the American Nurses Association help regulate standards for nursing practice. Organizations do provide socialization and keep data on members, but these functions don't have the same impact on nursing as do standards. Some nursing organizations may provide legal advice regarding job responsibilities, but they don't regulate the positions.

Taxonomic level: Knowledge

25. *Correct answer:* **D**

All registered nurses may perform the other options. Only nurses with advanced practice standing may write prescriptions in the states that allow this activity.

Taxonomic level: Application

26. *Correct answer:* **D**

The state in which a nurse practices controls her license. A nurse may have licensure in more that one state. The ANA, NLN, and AACN don't control an individual's license.

Taxonomic level: Knowledge

27. *Correct answer:* **D**

An MSN is earned after the bachelor's degree is acquired. A nursing master's usually prepares the nurse for a specialty or expanded role.

Taxonomic level: Knowledge

28. *Correct answer:* **B**

Functional nursing involves the assignment of tasks instead of patients to personnel (for example, one person would take all the blood pressures). In case management nursing, the patient is assigned to a health care provider responsible for coordinating the patient's care as long as the patient is within their service area, such as a hospital or health service. In team nursing, a group of caregivers is assigned a group of patients and the team decides how to care for the group. In total patient care nursing, a nurse is assigned a small group of patients and is responsible for all patient care within this group.

Taxonomic level: Knowledge

29. *Correct answer:* **C**

Medicare is the federal program that provides funding for people age 65 and older. Social Security is a federal program that provides a monthly payment if you have participated in the program throughout your working career. Medicaid, a federal program with no age restrictions, is designed for indigent individuals who need help with medical bills. HMOs are health maintenance organizations that one may join to receive health care.

Taxonomic level: Knowledge

30. *Correct answer:* **B**

DRG is the classifying system that determines the fees for hospitals and doctors who deliver care under the Medicare program. HMOs and PPOs are types of health insurance plans. CHAMPUS is a federal program to provide health care in the private sector for military personnel and their families.

Taxonomic level: Knowledge

31. *Correct answer:* **B**

Informed consent is in jeopardy if the patient is under the influence of narcotics or drugs that may alter thought processes. Informed consent papers may be signed the evening before, with family members present, and before all tests results are in the chart.

Taxonomic level: Knowledge

32. *Correct answer:* **A**

Battery is the unlawful use of force on a person, which could be alleged in this situation. "Challenge for cause" is a challenge based on a particular reason (such as bias) that a party (or his lawyer) can use to disqualify a juror. Comparative negligence refers to the determination of liability in which damages may be apportioned among multiple defendants. Negligence is the omission of an act that a prudent person would have performed. Slander is any words spoken with malice that are untrue and prejudicial to another person.

Taxonomic level: Knowledge

33. *Correct answer:* **A**

Good Samaritan laws are designed to protect health care professionals and others when they help people in emergency situations, unless care is given in a grossly negligent manner. Every state has its own Nurse Practice Act that is enforced by a governmental body, usually a State Board of Nursing. Minimum Standards are state standards that schools of nursing must abide by to stay approved so graduates of their schools are eligible to take the licensing examination to become registered nurses. The Occupational, Safety, and Health Act of 1970 (OSHA) sets legal standards in the United States in an effort to ensure safe and healthful working conditions.

Taxonomic level: Knowledge

34. *Correct answer:* **A**

This is an example of confidentiality. What will the doctor do with the information? Should the nurse be the one to tell the doctor? Conflict of interest is a conflict between an individual's private interests and his duties or actions as an official in a company or government. Negligence is the omission of an act that a prudent person would have performed. Slander is any words spoken with malice that are untrue and prejudicial to another person.

Taxonomic level: Knowledge

35. *Correct answer:* **C**

The definition of negligence is the omission of an act that a prudent person would have performed; in this case, the act is failure to raise the side rails. Malpractice is the proximate cause of injury or harm to a patient resulting from failure to act with the professional knowledge, experience, or skill that can be expected of others in the profession. Slander is any words spoken with malice that are untrue and prejudicial to another person. Battery is the unlawful use of force on a person.

Taxonomic level: Knowledge

36. *Correct answer:* **D**

Statutory laws are enacted by a legislative body in the United States and must be in keeping with the U.S. Constitution and the relevant state constitution. Administrative laws are laws that executive offices, such as the office of the President of the United States, or government agencies are empowered to make. Common laws arise from court decisions. Constitutional laws come from the U.S. Constitution and state constitutions as interpreted by the U.S. Supreme Court or the state's highest court, respectively.

Taxonomic level: Knowledge

37. *Correct answer:* **B**

Common laws arise from court decisions. Administrative laws are laws that executive offices, such as the office of the President of the United States, or administrative agencies are empowered to make. Statutory laws are enacted by a legislative body in the United States and must be in keeping with the U.S. Constitution and the specific state constitution. Constitutional laws come from the U.S. Constitution and state constitutions as interpreted by the U.S. Supreme Court or the state's highest court, respectively.

Taxonomic level: Knowledge

38. *Correct answer:* **A**

A concept or an idea is an abstract impression organized to form a theory. An outcome is the desired action of a goal. A process is the action phase that brings desired results. Philosophy is the study of wisdom, knowledge, and processes used to develop our perceptions of life.

Taxonomic level: Knowledge

39. *Correct answer:* **A**

A concept is an abstract impression organized to form a theory. Theories formed from concepts are used in various nursing situations. The other options don't apply.

Taxonomic level: Knowledge

40. *Correct answer:* **D**

The center of all theories is the person. Without people (for example, patients), a caring profession simply wouldn't exist.

Taxonomic level: Comprehension

41. *Correct answer:* **D**

Roy's theory is called the adaptation theory and states that the person is in constant interaction with a changing environment. Orem's theory is called self-care deficit theory and is based on the belief that the individual has a need for self-care actions. Neuman's model sees health as synonymous with wellness, when all parts are in harmony with the whole. Rogers' theory is an abstract theory of unitary human beings in which the person is an irreducible whole.

Taxonomic level: Knowledge

42. *Correct answer:* **C**

The self-care deficit theory of Dorothea Orem is based on the belief that the individual has a need for self-care actions. The transcultural theory is Madeline Leininger's, the adaptation theory is Sister Callista Roy's, and the behavioral system theory is Dorothy Johnson's.

Taxonomic level: Knowledge

43. *Correct answer:* **C**

Role theory was developed in the 1930s and 1940s by George Herbert Mead, Ralph Linton, and Jacob Moreno. They believed that social behavior, social process, and status play a significant part in how we perceive one another. General systems theory relates to the interdependency of systems. Adaptation theory states that a person is in constant interaction with a changing environment. Self-care deficit theory is based on the belief that an individual has a need for self-care actions.

Taxonomic level: Knowledge

44. *Correct answer:* **C**

Rogers' theory is the development of a science of humankind, incorporating the concepts of energy fields, openness, pattern, and organization. Johnson's theory is considered a complete behavioral system theory, Levine's theory is defined as the four conservation principles of nursing, and Peplau's theory is defined as an interpersonal process focusing on a patient with felt needs.

Taxonomic level: Knowledge

45. *Correct answer:* **A**

Betty Neuman's theory can be defined as conceptualized nursing as a total-person approach. Option B is Imogene King's theory, option C is Jean Watson's theory, and option D is Ida Orlando's theory.

Taxonomic level: Knowledge

46. *Correct answer:* **B**

Leininger's theory is known as the transcultural model. Henderson's theory is conceptualized nursing as assisting patients with 14 essential functions toward independence. Roy's theory is adaptation theory, and Orem's theory is the self-help theory.

Taxonomic level: Knowledge

47. *Correct answer:* **D**

The plaintiff is the person who files the lawsuit. The defendant is the person named in the lawsuit as the wrongdoer. The litigator is the attorney. Agitator isn't a formal role in the legal process.

Taxonomic level: Comprehension

48. *Correct answer:* **D**

Malpractice is defined as the proximate cause of injury or harm to a patient resulting from the failure to act with the professional knowledge, experience, or skill that can be expected of others in the profession. Libel is a false accusation — written, printed, or typed — that is made with malicious intent. Fraud is willful and purposeful misrepresentation that could cause harm to a person. Assault is a threat or attempt to make bodily contact with another person without the person's consent.

Taxonomic level: Knowledge

49. *Correct answer:* **B**

Drug and alcohol abuse is by far the most common reason that a nurse is placed on probation or has her license revoked. A nursing license can be revoked for a felony as well. Having a physical disability isn't a reason for license revocation.

Taxonomic level: Knowledge

50. *Correct answer:* **C**

Theft is the only one of the above situations that is a felony and would constitute a valid reason for revocation of a nursing license.

Taxonomic level: Knowledge

51. *Correct answer:* **B**

A nurse's responsibility is to know all medication that she administers. If the nurse administers the medication, she's responsible. If the nurse feels the dose isn't within normal limits, she should call the doctor who ordered the medication.

Taxonomic level: Application

52. *Correct answer:* **C**

A nurse is responsible for her own actions. Each employee has her own job description and is employed within that description. If the assignment for the nurse assistant falls within the job description, the nurse assistant is responsible for her own actions.

Taxonomic level: Application

53. *Correct answer:* **D**

Many governmental decisions, such as Medicare and diagnostic-related groups, affect nursing care and hospital stays. Doctors aren't trying to take over nursing duties. If nurses are needed for war, the military or other volunteers will take responsibility.

Taxonomic level: Knowledge

54. *Correct answer:* **A**

The *Federal Register* is published by the federal government and should include information for all U.S. constituents. Local newspapers and weekly news magazines are controlled by editors and often subtly reflect the writers' opinions. Senators' newsletters are the senators' opinions.

Taxonomic level: Knowledge

55. *Correct answer:* **C**

Negligence is the omission of an act that a prudent person would have performed. It is conduct that falls below the standard of care. Nonmaleficence means refraining from doing harm. Malpractice is defined as the proximate cause of injury or harm to a patient resulting from the failure to act using the professional knowledge, experience, or skill that can be expected of others in the profession. Libel is a false accusation—written, printed, or typed—that is made with malicious intent to defame a person's reputation.

Taxonomic level: Comprehension

56. *Correct answer:* **C**

Option C is the correct definition of fraud. Option A is the definition of libel, option B is the definition of slander, and option D is the definition of negligence.

Taxonomic level: Knowledge

57. *Correct answer:* **A**

The definition of fraud is the act of intentionally misleading or deceiving another person by any means. Altering a chart fits this definition. Malpractice is the proximate cause of injury or harm to a patient resulting from the failure to act using the professional knowledge, experience, or skill that can be expected of others in the profession. Libel is defined as a false accusation — written, printed, or typed — that is made with malicious intent. Slander is defined as any words spoken with malice that are untrue and prejudicial to another person.

Taxonomic level: Application

58. *Correct answer:* **C**

Depositions are legal testimony taken by attorneys and recorded outside the court proceedings. A tort is a civil wrong. Certified testimony is the statement of a witness that is usually made orally and given under oath such as in a court trial. A contract is an agreement that meets legal requirements between parties.

Taxonomic level: Knowledge

59. *Correct answer:* **B**

Civil laws govern contracts, ownership of property, and the practice of nursing, medicine, pharmacy, and dentistry. Administrative laws are laws that executive offices and administrative agencies are empowered to make. The common law derives from judicial decisions. Public laws are laws in which the government is directly involved.

Taxonomic level: Knowledge

60. *Correct answer:* **A**

The American Hospital Association published the Patient's Bill of Rights in 1973. The National League for Nursing had published a Bill of Rights in 1959, but failed to get information about it to the public.

Taxonomic level: Knowledge

61. *Correct answer:* **B**

A durable power of attorney gives a proxy the right to make treatment decisions for the person named in the document. A doctor's order can't transfer the patient's rights. A living will governs "life-sustaining" treatment only. A signed advance directive is a written document that provides directions concerning the provision of care when a person becomes unable to make his own treatment choices.

Taxonomic level: Comprehension

CHAPTER 4

COMMUNICATION, PATIENT TEACHING, AND LEADERSHIP

QUESTIONS

1. Nursing actions important in the primary prevention of violence and abuse include:

    A. helping participants discuss the problem and develop alternatives for dealing with the tension that could lead to violence.

    B. identifying "red flag" behaviors, including isolation and depression.

    C. emphasizing safety as a top priority.

    D. teaching patients the importance of respect and caring for family members.

2. Nursing actions important in the secondary prevention of violence and abuse include:

    A. helping participants discuss the problem and develop alternatives for dealing with the tension that could lead to violence.

    B. identifying "red flag" behaviors, including isolation and depression.

    C. emphasizing safety as a top priority.

    D. teaching patients the importance of respect and caring for family members.

3. The goal of risk appraisal and reduction is:

    A. empowerment of individuals to achieve health.

    B. improved quality of life.

    C. prevention or early detection of disease.

    D. reduced mortality.

4. Which statement is true about improving health behavior?

    A. Behavioral patterns of populations are related to habits of choice from actual or perceived limited resources.

    B. Health education affects behavioral patterns maximally, whether or not new health-promoting options are provided.

    C. Health status of populations isn't a function of the lack of, or excess of, health-sustaining resources.

    D. Social change isn't a reflection of a change in population behavior patterns.

5. It's generally agreed that community health nurses must:

    A. develop more community programs.

    B. improve their family assessment skills.

    C. rely on the biomedical model to guide their practice.

    D. shift their emphasis from illness to wellness.

6. Based on multiple referrals, the nurse determines that childhood injuries are increasing in the community in which she practices. The first step the nurse would take in developing an educational program is:

    A. assessing for a decrease in referrals following a pediatric safety class.

    B. assessing the strengths and needs of the community while identifying barriers to learning.

    C. choosing a health promotion or health belief model as a framework.

    D. developing and implementing a specific plan to decrease childhood injuries.

7. Phases of communication in the nurse-patient relationship are:

    A. assessment, counseling, and evaluation.

    B. assessment, planning, implementation, and evaluation.

    C. initiation, continuation, and termination.

    D. orientation, acceptance, and termination.

8. Potential barriers to learning in a rural farm community include:

    A. lack of trust in health care workers.

    B. low reading level.

    C. socioeconomic status.

    D. stereotypes of appropriate parenting techniques.

9. Long-term evaluation of a community education program will focus on:

    A. individual patient assessment of the program.

    B. long-term follow-up of program participants.

    C. monitoring of the number of pediatric accident victims at the local emergency department.

    D. monitoring of pediatric mortality statistics.

10. Which of the following choices is a potential or actual indicator of elder abuse?

    A. Fear of social interaction

    B. Fear of the caregiver

    C. Unexplained paranoia

    D. Unexplained weight loss

11. The goal of community health promotion for patients with coronary artery disease is:

    A. both prevention of complications and improved quality of life.

    B. improved quality of life.

    C. reduced incidence of fatal complications.

    D. reduced mortality from heart disease.

12. Which community prevention strategy is most helpful in preventing elder abuse?

    A. Church bingo classes for respite
    B. Home nursing care
    C. Providing information on discontinuing care at home
    D. Town watch

SITUATION: *Owen Roberts is a 50-year-old man admitted to the clinic with newly diagnosed diabetes.*

Questions 13 to 17 relate to this situation.

13. When doing patient teaching with Mr. Roberts, the nurse will assume the role of:

    A. doer.
    B. talker.
    C. teacher.
    D. thinker.

14. Mr. Roberts says, "I just want to quit." The most therapeutic nursing response would be:

    A. "Don't you think you're acting childish?"
    B. "Everyone wants to quit sometimes."
    C. "Tell me more about how you're feeling."
    D. "Well, you can't quit right now."

15. Therapeutic relationships are:

    A. helping relationships.
    B. nonreciprocal relationships.
    C. reciprocal relationships.
    D. social relationships.

16. The nurse determines that Mr. Roberts is benefiting from his teaching when:

    A. he admits that he has a problem.
    B. he cancels his teaching session to participate in an exercise class.
    C. he states: "I think I've got this diabetes licked."
    D. his presentation at his teaching session is a list of daily blood glucose levels over a week's time.

17. Which of the following choices accurately states the purpose of communication?

    A. To affect or influence others, one's physical environment, and oneself

    B. To discuss the patient's feelings

    C. To maintain initially defined boundaries

    D. To promote an attitude of acceptance

SITUATION: *Barbara Rollins is a 37-year-old woman with two young children. She has been admitted with severe depression.*

    Questions 18 to 21 relate to this situation.

18. During the initial phase of the nurse-patient relationship, the most helpful nursing intervention for Mrs. Rollins is:

    A. alleviating symptoms.

    B. assessing anxiety.

    C. providing sympathy.

    D. setting limits.

19. The nurse anticipates that Mrs. Rollins is attempting to achieve which of Erikson's developmental tasks?

    A. Generativity versus stagnation

    B. Identity versus role diffusion

    C. Initiative versus guilt

    D. Trust versus mistrust

20. The best way to promote communication with Mrs. Rollins is to:

    A. allow her to remain isolated in her room.

    B. ask for clarification and restate or paraphrase her statements.

    C. place strict time limits on her efforts at communication.

    D. tell her what you think is going on.

21. Techniques that may help Mrs. Rollins gain self-awareness include:

    A. discouraging comparison with other episodes in her life.

    B. discouraging her assessment of emotions.

    C. encouraging her to cry as frequently as needed.

    D. encouraging reflection.

SITUATION: *Justin Wright is a 15-year-old male who has been admitted to your unit after an attempted suicide.*

    Questions 22 to 25 relate to this situation.

22. The nurse anticipates that Justin is trying to achieve which of Erikson's developmental tasks?

  A. Identity versus role confusion
  B. Generativity versus stagnation
  C. Initiative versus guilt
  D. Trust versus mistrust

23. The nurse anticipates providing which of the following therapeutic modalities for Justin?

  A. Discouragement of focusing on body image
  B. Discouragement of striving for peer group approval
  C. Enforcement of strict limits
  D. Flexible enforcement of rules

24. The nurse assesses Justin for suicidal ideation, realizing that suicide is the _____ leading cause of death in this patient population.

  A. first
  B. second
  C. fourth
  D. tenth

25. The nurse would worry about a repeat attempt at suicide if Justin displayed which of the following behaviors?

  A. Severe withdrawal and a refusal to eat
  B. Sudden elation and chattiness
  C. Verbalization of plan to commit suicide
  D. Verbalization of plan not to commit suicide

SITUATION: *Michelle George is a 4-year-old girl admitted to the hospital with acute abdominal pain. She is recovering following an appendectomy.*

  Questions 26 to 29 relate to this situation.

26. Based on Michelle's age, the nurse anticipates that the child's experience of pain will be that she:

  A. attempts to delay procedures that are painful.
  B. believes she has pain because of something she did or thought.
  C. demonstrates high-pitched, loud crying.
  D. is able to precisely describe the pain.

27. The nurse anticipates that behavioral manifestations of pain in Michelle may include:

    A. regression to earlier behaviors.
    B. touching hurting body part.
    C. use of procrastination or bargaining to delay procedure.
    D. use of words to describe pain.

28. An appropriate assessment tool for the nurse to use when measuring pain in Michelle includes a:

    A. body outline and intensity scale.
    B. numeric rating scale from 1 to 100.
    C. pain descriptor list.
    D. scale with pictures.

29. In teaching Michelle about her illness, surgery, and plans for postoperative recovery, the nurse should incorporate knowledge that this patient's developmental stage is:

    A. generativity versus stagnation.
    B. identity versus role diffusion.
    C. initiative versus guilt.
    D. trust versus mistrust.

SITUATION: *Brenda Schaeffer is a 38-year-old housewife. She has been diagnosed with clinical depression.*

Questions 30 to 32 relate to this situation.

30. In a therapeutic relationship, the nurse assumes which role with Mrs. Schaeffer?

    A. Doer
    B. Friend
    C. Helper
    D. Listener

31. When beginning a therapeutic relationship with Mrs. Schaeffer, the first phase of nurse-patient interactions is known as the:

    A. helping phase.
    B. orientation phase.
    C. talking phase.
    D. working phase.

32. Which statement about patient touch is true?

    A. Most patients prefer not to be touched.
    B. Most patients prefer to be touched.
    C. Nurse–patient touching is an issue that requires sensitivity on the part of the nurse.
    D. Patients should never be touched.

SITUATION: *Bryan Frank is a 33-year-old man admitted with exacerbation of his ulcerative colitis. The nurse is performing an admission assessment.*

Questions 33 and 34 relate to this situation.

33. The nurse is assessing patient teaching needs regarding appropriate diet and lifestyle modifications for Mr. Frank. To develop an effective teaching plan, the nurse must solicit which of the following input from Mr. Frank?

    A. Details about his childhood phobias
    B. His feelings, beliefs, and attitudes about his chronic illness
    C. Information about his financial status
    D. Information about his relationship with his wife

34. Which choice best describes the phase of nurse-patient relationship the nurse is currently engaged in with Mr. Frank?

    A. Evaluation phase
    B. Orientation phase
    C. Termination phase
    D. Working phase

SITUATION: *Sheila Robinson is a 65-year-old woman who has been diagnosed with terminal breast cancer.*

Questions 35 to 37 relate to this situation.

35. Techniques that would be useful in eliciting information about how this patient is coping with her diagnosis include:

    A. asking closed questions.
    B. asking open-ended questions.
    C. soliciting information from the patient's family after interviewing the patient.
    D. soliciting information from the patient's family before interviewing the patient.

36. Objective data that may indicate anxiety in Ms. Robinson includes:

    A. decreased independence in daily living activities.
    B. increased distraction.
    C. increased independence in daily living activities.
    D. increased urinary frequency.

37. In arranging for home hospice care for Ms. Robinson, it's essential that the nurse assess which of the following items?

    A. The family's readiness for the patient to die
    B. The family's readiness to participate in patient care
    C. The patient's emotional stability
    D. The patient's readiness to die

SITUATION: *Peggy Del Sordo is a 53-year-old female newly diagnosed with hypertension. She's very confused about her treatment plan and her diagnosis.*

Questions 38 to 42 relate to this situation.

38. When providing Mrs. Del Sordo with teaching, the nurse must recognize which of the following statements about learning principles and patient teaching?

    A. The basic principle of learning is that it's an "other" activity.
    B. The patient/family system will modify essential teaching information provided.
    C. The patient is a whole person with psychological strengths and weaknesses.
    D. Transfer of learning usually takes place automatically.

39. When providing information to Mrs. Del Sordo about the purposes of her prescriptions, the nurse should:

    A. include the information on discharge planning sheets so the patient can review the information at her leisure.
    B. provide only bits and pieces of information at a time so as to not overwhelm the patient.
    C. provide the patient with a written teaching aid, which is reviewed with each dose of medication.
    D. wait for the family to be present to provide teaching.

40. The nurse anticipates which of the following possible barriers to learning for Mrs. Del Sordo?

A. Distraction
B. Fatigue
C. Fear and anxiety
D. Loneliness and anxiety

41. Which of the following choices is true regarding suitable teaching strategies for Mrs. Del Sordo?

A. Begin the teaching session with a detailed overview of material to be covered.
B. Long presentations are more effective than short presentations.
C. Personalizing standardized teaching information to this patient's particular educational and social background will prove most effective.
D. Teaching tools developed at a sixth-grade reading level are most appropriate.

42. The three domains of learning commonly used in patient teaching are:

A. cognitive, affective, and psychomotor.
B. cognitive, social, and psychomotor.
C. educational, affective, and psychomotor.
D. social, psychic, and educational.

SITUATION: *Andrew Clemmer is a 73-year-old man who has been admitted to the hospital for cataract surgery. He has a history of coronary artery disease.*

Questions 43 to 46 relate to this situation.

43. The nurse is assessing Mr. Clemmer's learning needs. Which statement is true regarding assessment of cognitive domain?

A. Long-term memory retains information long enough for processing.
B. Memory consists of three different phases.
C. Short-term memory permanently stores information.
D. The major intellectual process of learning involves memory.

44. In assessing Mr. Clemmer's affective domain, the nurse should ask questions about which of the following items?

A. Dietary preferences
B. Educational background
C. Preconceptions about heart disease
D. Religious affiliation

45. Which statement by Mr. Clemmer may indicate a lack of readiness to learn?

  A. "I can't wait to get out of the hospital to eat some real food."
  B. "I hope I can remember all the stuff you have taught me."
  C. "I'm feeling scared."
  D. "I'm going to have to change my eating habits."

46. Based on Mr. Clemmer's age, strategies for teaching him include:

  A. assessing for visual and auditory acuity.
  B. planning teaching sessions for first thing in the morning.
  C. providing plenty of details as needed.
  D. teaching while the patient is eating.

SITUATION: *Nancy Max is a 35-year-old woman with a history of infertility. She's admitted with complications following a miscarriage.*

Questions 47 to 50 relate to this situation.

47. Mrs. Max refuses to get out of bed. The nurse's most appropriate action is to:

  A. allow her to stay in bed until she feels ready to get up.
  B. refuse to give her breakfast until she gets up.
  C. tell her she must get up.
  D. tell her she will help her get up and get dressed.

48. It's determined that Mrs. Max is depressed. One of the first interventions for her should be aimed at:

  A. developing a good nursing care plan.
  B. developing a structured routine for her to follow.
  C. encouraging her to participate in activities of daily living.
  D. talking to her husband for cues to help her.

49. Which of the following support services may prove helpful to Mrs. Max?

  A. Nutritional support
  B. Occupational therapy
  C. Physical therapy
  D. Psychiatric liaison

50. When preparing Mrs. Max for discharge, the nurse is aware that which of the following issues is most important?

A. Applying for short-term disability benefits
B. Arranging for transportation
C. Referring the patient to community-based support groups
D. Referring the patient to nursing home care agency

SITUATION: *The nurse has organized a cancer support group that meets weekly. Questions 51 to 54 relate to this situation.*

51. Which statement best describes the nature of a group?

A. Groups promote the independence of individuals.
B. Groups share a unity of purpose.
C. Members of groups have different characteristics.
D. Members of groups have different goals.

52. As the group leader, which of the following best describes the nurse's role in the group?

A. Provide guidance and direction to the group.
B. Record and distribute minutes of weekly meetings.
C. Set dates and times of future meetings.
D. Serve as arbitrator during times of group discord.

53. A young member of the group is often angry and hostile. As the group leader, the nurse's best response to this member would be to:

A. allow other members of the group to confront the hostile member.
B. carefully confront the member following his outbursts.
C. carefully implement therapeutic silence following his outbursts.
D. change the group meeting times and dates without telling him.

54. A common understanding that is shared by the members of a support group is the:

A. acceptance that they all are going to die.
B. realization that each member isn't alone in experiencing cancer.
C. realization that some of the members are coping better than other members.
D. understanding that the stronger members are there to support the weaker members.

SITUATION: *Brenda Leedson is a 10-year-old girl who has scoliosis. She has been admitted for placement of a Harrington rod.*

Questions 55 to 58 relate to this situation.

55. According to Erikson's psychosocial stages, Brenda is in the stage of:

    A. autonomy versus shame and doubt.
    B. industry versus inferiority.
    C. initiative versus guilt.
    D. trust versus mistrust.

56. The nurse knows that the best tools for preoperative teaching for Brenda include:

    A. booklets with basic diagrams and simple written explanations.
    B. puppets.
    C. simple diagrams without explanations.
    D. simple three-dimensional models.

57. Barriers to learning for Brenda may include:

    A. fear and anxiety.
    B. fear of death.
    C. fear of mutilation.
    D. fear of pain.

58. The nurse anticipates that Brenda may require which of the following community resources at discharge?

    A. Family support group
    B. Outpatient psychiatric care
    C. Physical therapy and rehabilitation
    D. Social service

SITUATION: *Brandon Wilson is a 28-month-old boy admitted to the hospital with acute croup.*

Questions 59 to 62 relate to this situation.

59. Which of the following would be most threatening to Brandon's autonomy?

    A. Complete bed rest
    B. Frequent visits by friends and family
    C. Participation in playroom activities with other children
    D. Riding to the X-ray department in a wheelchair

60. The most appropriate role for Brandon's mother to assume during his hospitalization is that of:

   A. primary caretaker.
   B. primary companion.
   C. secondary caretaker.
   D. secondary companion.

61. The nurse's admission assessment on Brandon reveals that he's in the lowest 10th percentile for his weight. His mother admits to having lost her job a month ago and states that she's experiencing financial difficulty. Upon discharge, the most important community resource she should be referred to is:

   A. community education classes.
   B. Medicaid.
   C. a well-child clinic.
   D. a Women, Infants, and Children (WIC) program.

62. The most appropriate tool for use in evaluating the appropriateness of Brandon's psychosocial development is the:

   A. Brazelton Neonatal Behavioral Assessment Scale.
   B. Denver Articulation Screening Examination.
   C. Early Language Milestone Scale.
   D. Revised Denver Prescreening Developmental Questionnaire.

SITUATION: *Salvatore Manzolilla is a 59-year-old man with severe osteoarthritis. He was hospitalized for hip replacement surgery 1 week ago. He's discharged to home with visiting nurses and home physical therapy. On the nurse's first home visit, she's performing a comprehensive assessment.*

Questions 63 to 70 relate to this situation.

63. On arrival, the nurse notes that Mr. Manzolilla is frequently grimacing and repositioning himself in a chair with a lot of difficulty. The best intervention the nurse can make at this point would be to:

   A. administer pain medication to alleviate discomfort.
   B. continue the interview as a means of distraction.
   C. involve the patient's primary caregiver in the discussion because the patient is distracted.
   D. relocate the patient from a chair to a bed to promote comfort.

64. Which home assessment finding warrants health promotion teaching from the nurse?

     A. A bathroom with grab bars for tub and toilet

     B. Items stored in the kitchen so that reaching up and bending down aren't necessary

     C. Many small, unsecured area rugs

     D. Sufficient stairwell lighting, with switches at the top and bottom of the stairs

65. To prevent problems with immobility in Mr. Manzolilla, the nurse should assess for:

     A. adequate nutrition and exercise.

     B. depression.

     C. fecal or urinary incontinence.

     D. iatrogenic drug reactions.

66. The nurse assesses for depression in this homebound patient. Clinical signs that alert the nurse that Mr. Manzolilla may be depressed include:

     A. decreased concentration and appetite.

     B. denial regarding his physical limitations.

     C. frequent complaints of postoperative discomfort.

     D. increased interest in activities.

67. The community health nurse is aware that the most common acute condition experienced by elderly people is:

     A. heart attack.

     B. respiratory tract infection.

     C. stroke.

     D. urinary tract infection.

68. The most feared chronic condition among the aging population is:

     A. cancer.

     B. dementia.

     C. heart disease.

     D. pulmonary disease.

69. The best overall description of a community health nurse's role is:

     A. prevention of disease and management of health care referrals.

     B. prevention of disease and promotion and maintenance of health.

     C. promotion and maintenance of health and management of health care referrals.

     D. promotion and maintenance of health and prevention of accidents.

70. The goal of health care for elderly people in the community is:

    A. optimizing their general health and functional capabilities.
    B. optimizing their socioeconomic status and reducing costs.
    C. preserving functional status.
    D. preserving physical status.

SITUATION: *The community health nurse serves a community whose residents are primarily from a low-income senior housing complex.*

    Questions 71 to 75 relate to this situation.

71. Health promotional activities that serve as primary prevention for this population include:

    A. assessing for visual or hearing impairments.
    B. intervening with residents who have suicidal ideations.
    C. offering pneumovax and influenza vaccinations.
    D. screening for breast or prostate cancer.

72. The nurse sponsors a health promotion day for seniors in the complex. Appropriate activities to aid in secondary prevention include:

    A. information on the importance of aspirin in patients susceptible to cardiovascular disease.
    B. information on the importance of hormone replacement therapy for postmenopausal women.
    C. instruction on accident prevention in the bathroom.
    D. screening for hypertension and hyperlipidemia.

73. The nurse organizes a support group for elderly people in the complex who have lost a spouse. This is an example of:

    A. primary prevention.
    B. secondary prevention.
    C. social directing.
    D. tertiary prevention.

74. The nurse is planning health promotional activities for these patients. Approximately what percentage of elderly people in the community express sadness or have a clinically diagnosable depression?

    A. 5% to 10%
    B. 10% to 20%
    C. 20% to 30%
    D. 30% to 40%

75. During home care visits, the nurse should assess for signs of cognitive impairments. What percentage of those over age 75 have clinically detectable cognitive impairment?

    A. 5%
    B. 10%
    C. 15%
    D. 20%

SITUATION: *Suzann DiCicco is a 35-year-old woman receiving evaluation in the emergency department for facial injuries and a fractured left ulna. A chest X-ray reveals multiple rib fractures in various stages of healing. The nurse suspects spousal abuse.*

Questions 76 to 79 relate to this situation.

76. What is the best question to ask regarding Mrs. DiCicco's facial injuries, including a blackened eye and sphenoid sinus fractures?

    A. "What happened to your face?"
    B. "What happened to you?"
    C. "Did anybody else see what happened?"
    D. "Who hit you?"

77. Mrs. DiCicco states that she blames herself for her injuries. The nurse understand that Mrs. DiCicco probably suffers from:

    A. a history of child abuse.
    B. chronic head injury.
    C. low self-esteem and depression.
    D. substance abuse.

78. When discharging Mrs. DiCicco to her home, the nurse should provide her with information for which of the following services?

    A. Individual counseling
    B. Marital counseling
    C. Marital counseling and a women's shelter
    D. A women's shelter

79. Mrs. DiCicco's batterer is legally charged in her abuse. Which of the following community resources is associated with the best recovery rates for batterers?

    A. Court-mandated programs that focus on individual therapy
    B. Court-mandated programs that focus on men's underlying values about women
    C. Prison programs that focus on violence
    D. Prison programs with group therapy sessions

SITUATION: *Holly Wilkerson is a 31-year-old woman who's been admitted to the hospital for observation after she attempted suicide.*

Questions 80 and 81 relate to this situation.

80. The nurse includes a psychosocial assessment that includes Ms. Wilkerson's marital relationship, based on the nurse's knowledge that:

    A. spouses are often the first to be aware of a potential suicide.
    B. spouses may be able to intervene in future suicide attempts.
    C. suicide attempts may adversely affect the marital relationship.
    D. the number one risk factor for suicide in adult women is spousal abuse.

81. Psychiatric follow-up for Ms. Wilkerson is essential because she:

    A. is definitely depressed.
    B. is probably angry that she's still alive.
    C. may try to commit suicide again.
    D. obviously hates her life.

SITUATION: *Kadim Jabar is a 16-year-old black male, discharged to home after treatment of a gunshot wound to the arm.*

Questions 82 and 83 relate to this situation.

82. The nurse includes Kadim's mother in discharge teaching. When addressing the issue of violence, the nurse informs her that homicide is the _____ leading cause of death among young black men and women.

    A. first
    B. second
    C. third
    D. fourth

83. It's revealed that Kadim's wound occurred as the result of a family argument. The nurse notifies social services and arranges for a community assessment and follow-up based on her knowledge that:

    A. abuse within families isn't related to the incidence of homicide.
    B. abuse within a family is usually followed by homicide.
    C. homicide within families almost never occurs.
    D. homicide within families is usually preceded by the abuse of a family member.

ANSWER SHEET

	A B C D		A B C D		A B C D		A B C D
1	○ ○ ○ ○	22	○ ○ ○ ○	43	○ ○ ○ ○	64	○ ○ ○ ○
2	○ ○ ○ ○	23	○ ○ ○ ○	44	○ ○ ○ ○	65	○ ○ ○ ○
3	○ ○ ○ ○	24	○ ○ ○ ○	45	○ ○ ○ ○	66	○ ○ ○ ○
4	○ ○ ○ ○	25	○ ○ ○ ○	46	○ ○ ○ ○	67	○ ○ ○ ○
5	○ ○ ○ ○	26	○ ○ ○ ○	47	○ ○ ○ ○	68	○ ○ ○ ○
6	○ ○ ○ ○	27	○ ○ ○ ○	48	○ ○ ○ ○	69	○ ○ ○ ○
7	○ ○ ○ ○	28	○ ○ ○ ○	49	○ ○ ○ ○	70	○ ○ ○ ○
8	○ ○ ○ ○	29	○ ○ ○ ○	50	○ ○ ○ ○	71	○ ○ ○ ○
9	○ ○ ○ ○	30	○ ○ ○ ○	51	○ ○ ○ ○	72	○ ○ ○ ○
10	○ ○ ○ ○	31	○ ○ ○ ○	52	○ ○ ○ ○	73	○ ○ ○ ○
11	○ ○ ○ ○	32	○ ○ ○ ○	53	○ ○ ○ ○	74	○ ○ ○ ○
12	○ ○ ○ ○	33	○ ○ ○ ○	54	○ ○ ○ ○	75	○ ○ ○ ○
13	○ ○ ○ ○	34	○ ○ ○ ○	55	○ ○ ○ ○	76	○ ○ ○ ○
14	○ ○ ○ ○	35	○ ○ ○ ○	56	○ ○ ○ ○	77	○ ○ ○ ○
15	○ ○ ○ ○	36	○ ○ ○ ○	57	○ ○ ○ ○	78	○ ○ ○ ○
16	○ ○ ○ ○	37	○ ○ ○ ○	58	○ ○ ○ ○	79	○ ○ ○ ○
17	○ ○ ○ ○	38	○ ○ ○ ○	59	○ ○ ○ ○	80	○ ○ ○ ○
18	○ ○ ○ ○	39	○ ○ ○ ○	60	○ ○ ○ ○	81	○ ○ ○ ○
19	○ ○ ○ ○	40	○ ○ ○ ○	61	○ ○ ○ ○	82	○ ○ ○ ○
20	○ ○ ○ ○	41	○ ○ ○ ○	62	○ ○ ○ ○	83	○ ○ ○ ○
21	○ ○ ○ ○	42	○ ○ ○ ○	63	○ ○ ○ ○		

ANSWERS AND RATIONALES

1. *Correct answer:* **B**

Primary prevention encompasses the assessment of risk factors—in this case, risk factors leading to violence and abuse—which include depression and social isolation.

> *Nursing process step:* Assessment
> *Client needs category:* Safe, effective care environment
> *Client needs subcategory:* Management of care
> *Taxonomic level:* Application

2. *Correct answer:* **A**

Nursing measures in the secondary prevention of violence and abuse are intended to reduce the further incidence of violence and abuse once those behaviors have occurred. Nursing intervention in these situations is aimed at helping the participants, notably the victim, discuss the problem, seek alternative actions, and move to a safe haven if needed.

> *Nursing process step:* Assessment
> *Client needs category:* Safe, effective care environment
> *Client needs subcategory:* Management of care
> *Taxonomic level:* Analysis

3. *Correct answer:* **C**

The goal of risk appraisal and reduction is the prevention or early detection of disease.

> *Nursing process step:* N/A
> *Client needs category:* Physiological integrity
> *Client needs subcategory:* Reduction of risk potential
> *Taxonomic level:* Comprehension

4. *Correct answer:* **A**

To improve health behavior, people need to be provided with a range of resources and options. Health education does not always have a maximal impact on behavior patterns. Social change is indeed a reflection of change in population behavior patterns, and the health status of populations is in part a function of the lack of, or excess of, health-sustaining resources.

> *Nursing process step:* N/A
> *Client needs category:* Psychosocial integrity
> *Client needs subcategory:* Coping and adaptation
> *Taxonomic level:* Analysis

5. *Correct answer:* **D**

An understanding of the need to shift the emphasis from illness to wellness is increasing in community health nursing.

> *Nursing process step:* N/A
> *Client needs category:* Health promotion and maintenance
> *Client needs subcategory:* Prevention and early detection of disease
> *Taxonomic level:* Comprehension

6. *Correct answer:* **B**

Following the identification of a learning need, the first step is to assess the strengths and needs of the community while identifying barriers to learning.

> *Nursing process step:* Planning
> *Client needs category:* Safe, effective care management
> *Client needs subcategory:* Management of care
> *Taxonomic level:* Analysis

7. *Correct answer:* **C**

Phases in the nurse-patient relationship are the initiation/orientation phase, the continuation/active working phase, and the termination phase.

> *Nursing process step:* N/A
> *Client needs category:* Safe, effective care environment
> *Client needs subcategory:* Management of care
> *Taxonomic level:* Knowledge

8. *Correct answer:* **D**

Barriers to learning include preconceived cultural and ethnic norms of behavior regardless of community location. Although reading level, socioeconomic status, and lack of trust in health care workers would require revision of educational techniques, they would not necessarily present a barrier to learning.

> *Nursing process step:* Assessment
> *Client needs category:* Health promotion and maintenance
> *Client needs subcategory:* Growth and development through the life span
> *Taxonomic level:* Application

9. *Correct answer:* **C**

Long-term evaluation of a community education program focuses on the assessment of community statistics over time, such as the number of accident victims.

Nursing process step: Evaluation
Client needs category: Safe, effective care environment
Client needs subcategory: Management of care
Taxonomic level: Application

10. *Correct answer:* **B**

Fear of the caregiver is an indicator of potential or actual elder abuse. The other listed symptoms aren't necessarily indicative of elder abuse.

Nursing process step: N/A
Client needs category: Psychosocial integrity
Client needs subcategory: Psychosocial adaptation
Taxonomic level: Comprehension

11. *Correct answer:* **A**

Community health promotion involves both the promotion of a positive quality of life as well as the care and prevention of illness.

Nursing process step: N/A
Client needs category: Psychosocial integrity
Client needs subcategory: Coping and adaptation
Taxonomic level: Application

12. *Correct answer:* **C**

Developing new ways to help family caregivers, including helping them with decisions about discontinuing care at home, is helpful in preventing elder abuse. The other options aren't particularly effective in preventing elder abuse.

Nursing process step: N/A
Client needs category: Safe, effective care environment
Client needs subcategory: Management of care
Taxonomic level: Application

13. *Correct answer:* **C**

The nurse will assume the role of teacher in this therapeutic relationship. The other roles are inappropriate in this situation.

Nursing process step: Implementation
Client needs category: Physiological integrity
Client needs subcategory: Reduction of risk potential
Taxonomic level: Application

14. *Correct answer:* **C**

The therapeutic relationship should remain focused on eliciting the patient's feelings, thoughts, and values. Therefore, the best response is to ask the patient to expand on his comment, which helps him to verbalize his feelings. Option B reflects the patient's statement but doesn't encourage him to further verbalize his feelings; the other options are incorrect.

> *Nursing process step:* Implementation
> *Client needs category:* Psychosocial integrity
> *Client needs subcategory:* Coping and adaptation
> *Taxonomic level:* Application

15. *Correct answer:* **A**

A therapeutic relationship is a helping relationship that is not necessarily reciprocal. The nurse and patient are viewed as distinct, unique systems that intersect on a common ground. A therapeutic relationship isn't social. Social relationships tend to be reciprocal, with both persons sharing feelings, beliefs, and opinions with each other.

> *Nursing process step:* N/A
> *Client needs category:* Psychosocial integrity
> *Client needs subcategory:* Coping and adaptation
> *Taxonomic level:* Comprehension

16. *Correct answer:* **D**

A patient who's committed to a nurse-patient relationship and is benefiting from teaching will attend teaching sessions as contracted. Additionally, this patient is actively engaged in the process of managing his disease. The other answers don't indicate any particular benefit from the nurse-patient relationship.

> *Nursing process step:* Evaluation
> *Client needs category:* Psychosocial integrity
> *Client needs subcategory:* Coping and adaptation
> *Taxonomic level:* Analysis

17. *Correct answer:* **A**

The purposes of communication include transferring ideas from one to another, creating meaning through the process; and influencing others, one's physical environment, and oneself. Other purposes include reducing uncertainty, acting effectively, and defending or strengthening one's ego.

> *Nursing process step:* N/A
> *Client needs category:* Safe, effective care environment
> *Client needs subcategory:* Management of care
> *Taxonomic level:* Comprehension

18. *Correct answer:* **B**

During the initial phase of the nurse-patient relationship in this situation, tasks include establishing the boundaries of the relationship, identifying problems, assessing anxiety levels, and identifying expectations. All other responses aren't part of this phase of the relationship.

> *Nursing process step:* Evaluation
> *Client needs category:* Psychosocial integrity
> *Client needs subcategory:* Coping and adaptation
> *Taxonomic level:* Comprehension

19. *Correct answer:* **A**

According to Erik Erikson, the developmental task of adulthood is that of generativity versus stagnation or self-absorption. Identity versus role diffusion occurs during puberty and adolescence, initiative versus guilt occurs during the preschool years, and trust versus mistrust occurs during infancy.

> *Nursing process step:* Assessment
> *Client needs category:* Health promotion and maintenance
> *Client needs subcategory:* Growth and development through the life span
> *Taxonomic level:* Comprehension

20. *Correct answer:* **B**

Asking for clarification and restating or paraphrasing the patient's statements are techniques used to further elicit and clarify the patient's feelings. The other options aren't recommended methods for promoting communication.

> *Nursing process step:* Planning
> *Client needs category:* Psychosocial integrity
> *Client needs subcategory:* Coping and adaptation
> *Taxonomic level:* Knowledge

21. *Correct answer:* **D**

Encouraging the patient to reflect enables her to think about events and feelings and reach a conclusion. This may help her gain confidence in making assessments and decisions, and it may encourage self-reliance.

> *Nursing process step:* N/A
> *Client needs category:* Psychosocial integrity
> *Client needs subcategory:* Coping and adaptation
> *Taxonomic level:* Application

22. *Correct answer:* **A**

The young adolescent is typically struggling with the task of identity versus role confusion. Generativity versus stagnation occurs in middle adulthood, initiative versus guilt occurs in preschoolers, and trust versus mistrust occurs in infancy.

> *Nursing process step:* Assessment
> *Client needs category:* Health promotion and maintenance
> *Client needs subcategory:* Growth and development through the life span
> *Taxonomic level:* Comprehension

23. *Correct answer:* **D**

Guidelines for dealing with adolescents include providing looser limits to ensure security, showing flexibility in adjusting to mood swings, and being calm and consistent with flexible enforcement of rules.

> *Nursing process step:* Planning
> *Client needs category:* Psychosocial integrity
> *Client needs subcategory:* Psychosocial adaptation
> *Taxonomic level:* Application

24. *Correct answer:* **C**

Suicide is the fourth leading cause of death in the 15- to 40-year-old group.

> *Nursing process step:* Assessment
> *Client needs category:* Psychosocial integrity
> *Client needs subcategory:* Psychosocial adaptation
> *Taxonomic level:* Knowledge

25. *Correct answer:* **C**

Patients with a continued desire to commit suicide will usually verbalize it when questioned. The danger period for repeat suicide attempts occurs when the depression just begins to lift.

> *Nursing process step:* Analysis
> *Client needs category:* Psychosocial integrity
> *Client needs subcategory:* Psychosocial adaptation
> *Taxonomic level:* Application

26. *Correct answer:* **B**

Preschoolers may believe they're being punished with pain for something they did or thought. They're egocentric and tend to relate to the here and now; they're unable to relate discomfort to any positive outcome.

Nursing process step: Assessment
Client needs category: Health promotion and maintenance
Client needs subcategory: Growth and development through the life span
Taxonomic level: Application

27. *Correct answer:* **A**

Preschoolers frequently regress to earlier behaviors, such as loss of bowel or bladder control, when stressed or in pain. Touching the hurting body part is more often seen in toddlers and preschoolers. The other options aren't behavioral responses to pain by children in this age-group.

Nursing process step: Analysis
Client needs category: Psychosocial integrity
Client needs subcategory: Coping and adaptation
Taxonomic level: Application

28. *Correct answer:* **D**

The most appropriate pain evaluation tool for use with preschoolers is the use of a photographic scale depicting facial expressions of increasing levels of pain. The other tools are too sophisticated for use with this age-group.

Nursing process step: Planning
Client needs category: Psychosocial integrity
Client needs subcategory: Coping and adaptation
Taxonomic level: Application

29. *Correct answer:* **C**

Preschoolers are engaged in Erikson's developmental stage of initiative versus guilt.

Nursing process step: Planning
Client needs category: Health promotion and maintenance
Client needs subcategory: Growth and development through the life span
Taxonomic level: Analysis

30. *Correct answer:* **C**

A therapeutic relationship is a helping relationship.

Nursing process step: Assessment
Client needs category: Psychosocial integrity
Client needs subcategory: Coping and adaptation
Taxonomic level: Knowledge

31. *Correct answer:* **B**

During the orientation phase of the therapeutic relationship, the nurse and patient make an agreement that they will be working together to solve one or more of the patient's problems.

> *Nursing process step:* Implementation
> *Client needs category:* Psychosocial integrity
> *Client needs subcategory:* Coping and adaptation
> *Taxonomic level:* Knowledge

32. *Correct answer:* **C**

Touch has many meanings to patients. Although many patients are eager for human touch, others may perceive touch as human-boundary violation. Therefore, touching requires sensitivity on the part of the nurse.

> *Nursing process step:* Evaluation
> *Client needs category:* Psychosocial integrity
> *Client needs subcategory:* Coping and adaptation
> *Taxonomic level:* Comprehension

33. *Correct answer:* **B**

Assessment data should include information regarding the patient's feelings, beliefs, and attitudes about his illness. Although the other options are partially correct, option B is the best answer.

> *Nursing process step:* Planning
> *Client needs category:* Psychosocial integrity
> *Client needs subcategory:* Coping and adaptation
> *Taxonomic level:* Application

34. *Correct answer:* **D**

In the working phase of the nurse-patient relationship, the nurse actively participates in further patient assessment and in exploring and examining the data found.

> *Nursing process step:* Evaluation
> *Client needs category:* Psychosocial integrity
> *Client needs subcategory:* Coping and adaptation
> *Taxonomic level:* Comprehension

35. *Correct answer:* **B**

Asking open-ended questions that require a complex response is the best way for nurses to elicit information from patients. Closed questions that are answered with "yes" or "no" typically provide only limited information. Although seeking information from the patient's family may seem worthwhile, it isn't necessarily the best way to gain patient information because of confidentiality issues.

> *Nursing process step:* Planning
> *Client needs category:* Psychosocial integrity
> *Client needs subcategory:* Coping and adaptation
> *Taxonomic level:* Application

36. *Correct answer:* **D**

Objective data is measurable. Increased urinary frequency is measurable data that is related to anxiety. Independence of daily living and distraction aren't measurable quantities.

> *Nursing process step:* Assessment
> *Client needs category:* Psychosocial integrity
> *Client needs subcategory:* Coping and adaptation
> *Taxonomic level:* Knowledge

37. *Correct answer:* **B**

It's essential to assess the readiness of the patient's family to participate in patient care when discharging a patient to home with hospice care. Although the other options address important issues, option B is the best response.

> *Nursing process step:* Assessment
> *Client needs category:* Psychosocial integrity
> *Client needs subcategory:* Coping and adaptation
> *Taxonomic level:* Application

38. *Correct answer:* **C**

The nurse must recognize each learner as a whole person with psychological strengths and weaknesses as well as health care experiences that play a role in her learning. Information provided should be modified to fit the needs of the patient and family system by the teacher, not the learner. Learning is a "self" activity, and the transfer of information is purposeful, not automatic.

> *Nursing process step:* Planning
> *Client needs category:* Psychosocial integrity
> *Client needs subcategory:* Psychosocial adaptation
> *Taxonomic level:* Analysis

39. *Correct answer:* **C**

A written teaching aid, which is reviewed with each dose of medication, is the best way to engage the patient and keep her involved in the learning process.

> *Nursing process step:* Planning
> *Client needs category:* Psychosocial integrity
> *Client needs subcategory:* Coping and adaptation
> *Taxonomic level:* Comprehension

40. *Correct answer:* **C**

Fear and anxiety are common in those patients facing a new diagnosis and an introduction to new medications. Although the other responses may be present, fear and anxiety tend to be the most prevalent.

> *Nursing process step:* Planning
> *Client needs category:* Psychosocial integrity
> *Client needs subcategory:* Coping and adaptation
> *Taxonomic level:* Comprehension

41. *Correct answer:* **C**

Personalizing standardized teaching information to adapt it to each situation and setting, and to the capabilities of the patient and family system fosters the learner's self-activity in the educative process.

> *Nursing process step:* Planning
> *Client needs category:* Psychosocial integrity
> *Client needs subcategory:* Coping and adaptation
> *Taxonomic level:* Comprehension

42. *Correct answer:* **A**

The three domains of learning commonly used in patient teaching are cognitive, affective, and psychomotor.

> *Nursing process step:* N/A
> *Client needs category:* Psychosocial integrity
> *Client needs subcategory:* Coping and adaptation
> *Taxonomic level:* Knowledge

43. *Correct answer:* **D**

Learning involves memory, which consists of short-term (processing) and long-term (storage) phases.

Nursing process step: Assessment
Client needs category: Psychosocial integrity
Client needs subcategory: Coping and adaptation
Taxonomic level: Knowledge

44. *Correct answer:* C

The affective domain deals with a person's thoughts, feelings, and attitudes, including any preconceptions the person may have regarding his health or diagnoses.

Nursing process step: Implementation
Client needs category: Psychosocial integrity
Client needs subcategory: Coping and adaptation
Taxonomic level: Knowledge

45. *Correct answer:* A

The patient's readiness to learn is influenced by several factors, including maturity, levels of anxiety and fear, and current educational needs. It's appropriate for the patient to voice these feelings. Barriers to learning include denial, severe anxiety, and fear.

Nursing process step: N/A
Client needs category: Psychosocial integrity
Client needs subcategory: Coping and adaptation
Taxonomic level: Application

46. *Correct answer:* A

Strategies for teaching older adults include assessing for visual and auditory acuity, as well as assessing for fatigue and hunger. Teaching should take place at a time when the person is able to concentrate, not during meals or immediately following sleep. Explanations should be as simple and as concrete as possible.

Nursing process step: N/A
Client needs category: Psychosocial integrity
Client needs subcategory: Psychosocial adaptation
Taxonomic level: Application

47. *Correct answer:* D

The nurse should be positive, definite, and specific about expectations. Physically assisting the patient to get up and be mobile is the best answer. The nurse should not give resistant patients choices or try to convince them to get up.

Nursing process step: Implementation
Client needs category: Psychosocial integrity
Client needs subcategory: Psychosocial adaptation
Taxonomic level: Knowledge

48. *Correct answer:* **B**

The priority for this patient is to get her mobilized. Providing a structured plan of activities for this patient to follow will help lift her mood and provide focus. The other options don't indicate appropriate interventions in this case.

> *Nursing process step:* Implementation
> *Client needs category:* Psychosocial integrity
> *Client needs subcategory:* Coping and adaptation
> *Taxonomic level:* Knowledge

49. *Correct answer:* **D**

The psychiatric support service should be consulted when developing a plan of care for this patient.

> *Nursing process step:* N/A
> *Client needs category:* Psychosocial integrity
> *Client needs subcategory:* Psychosocial adaptation
> *Taxonomic level:* Knowledge

50. *Correct answer:* **C**

Referral to community-based support groups will provide Mrs. Max with the continued support she'll require following her hospitalization.

> *Nursing process step:* Evaluation
> *Client needs category:* Safe, effective care environment
> *Client needs subcategory:* Management of care
> *Taxonomic level:* Application

51. *Correct answer:* **B**

The most important characteristic of a group is the concept of shared purpose. Groups also promote the interdependence of individuals and have members who share common goals and characteristics.

> *Nursing process step:* N/A
> *Client needs category:* Psychosocial integrity
> *Client needs subcategory:* Psychosocial adaptation
> *Taxonomic level:* Knowledge

52. *Correct answer:* **A**

The group leader provides guidance and direction to the group. Although the leader may also assume some of the other responsibilities listed above, option A is the best answer.

Nursing process step: Analysis
Client needs category: Safe, effective care environment
Client needs subcategory: Management of care
Taxonomic level: Comprehension

53. *Correct answer:* **C**

As the group leader, it's best for the nurse to directly confront, in a nonthreatening manner, the group member who's acting out. Uncomfortable as the process of confrontation is for group members and their leader, the failure to resolve important emotional responses to the group process can negatively affect outcomes.

Nursing process step: Implementation
Client needs category: Psychosocial integrity
Client needs subcategory: Psychosocial adaptation
Taxonomic level: Application

54. *Correct answer:* **B**

Although all of the statements are true to some extent, the best answer is option B, the realization that each member isn't alone in experiencing cancer.

Nursing process step: N/A
Client needs category: Psychosocial integrity
Client needs subcategory: Psychosocial adaptation
Taxonomic level: Comprehension

55. *Correct answer:* **B**

A 10-year-old is engaged in the psychosocial stage of industry versus inferiority. She's learning to gain attention by her accomplishments, to explore things, and to relate to groups of her own gender.

Nursing process step: N/A
Client needs category: Health promotion and maintenance
Client needs subcategory: Growth and development through the life span
Taxonomic level: Comprehension

56. *Correct answer:* **A**

Brenda is in Jean Piaget's formal operational stage of cognitive development and is capable of inductive and deductive logic. The nurse should anticipate that she will be able to most effectively use a teaching tool with both diagrams and written explanations.

Nursing process step: Evaluation
Client needs category: Physiological integrity
Client needs subcategory: Reduction of risk potential
Taxonomic level: Analysis

57. *Correct answer:* **A**

Preschoolers and early-school-age children typically suffer from fears of dying, mutilation, and pain. The older preadolescent is more likely to manifest symptoms of anxiety and fear.

> *Nursing process step:* N/A
> *Client needs category:* Psychosocial integrity
> *Client needs subcategory:* Coping and adaptation
> *Taxonomic level:* Analysis

58. *Correct answer:* **C**

Brenda will require physical therapy and rehabilitation to ensure continued recovery and mobility. Although the other choices may be needed, option C is the best answer.

> *Nursing process step:* Planning
> *Client needs category:* Safe, effective care environment
> *Client needs subcategory:* Management of care
> *Taxonomic level:* Comprehension

59. *Correct answer:* **A**

One of the greatest threats to a hospitalized toddler's autonomy is complete bed rest. He's just beginning to assert his independence, is very active, and doesn't want to be kept in bed.

> *Nursing process step:* N/A
> *Client needs category:* Psychosocial adaptation
> *Client needs subcategory:* Coping and adaptation
> *Taxonomic level:* Comprehension

60. *Correct answer:* **A**

A hospitalized toddler responds best to care provided by or in the presence of his primary caretaker, usually his mother.

> *Nursing process step:* N/A
> *Client needs category:* Psychosocial adaptation
> *Client needs subcategory:* Coping and adaptation
> *Taxonomic level:* Knowledge

61. *Correct answer:* **D**

Although community education classes and Medicaid may also be appropriate for this patient, the WIC program specifically will provide this parent with the resources necessary to buy food for her family. A well-child clinic would be no help at all.

Nursing process step: N/A
Client needs category: Safe, effective care environment
Client needs subcategory: Management of care
Taxonomic level: Analysis

62. *Correct answer:* **D**

For children in this age-group, either the Revised Denver Prescreening Developmental Questionnaire or the Developmental Profile II are used as developmental
screens. The Brazelton Neonatal Behavioral Assessment Scale is used to assess infant behavior in early neonates, the Denver Articulation Screening Examination
is for 2- to 6-year-olds, and the Early Language Milestone Scale is used to assess
speech and language rather than developmental status in this age-group.

Nursing process step: N/A
Client needs category: Health promotion and maintenance
Client needs subcategory: Growth and development through the life span
Taxonomic level: Application

63. *Correct answer:* **A**

The best way to gain the cooperation of the patient is to ensure his comfort, regardless of his positioning.

Nursing process step: Implementation
Client needs category: Physiological integrity
Client needs subcategory: Basic care and comfort
Taxonomic level: Application

64. *Correct answer:* **C**

The presence of unsecured area rugs poses a hazard in all homes, particularly in
one with a resident at high risk for falls.

Nursing process step: Evaluation
Client needs category: Safe, effective care environment
Client needs subcategory: Safety and infection control
Taxonomic level: Comprehension

65. *Correct answer:* **A**

The most important preventative action for immobility in elderly patients is to
ensure adequate nutrition and exercise, in addition to the provision of assistive
aids.

Nursing process step: Planning
Client needs category: Physiological integrity
Client needs subcategory: Reduction of risk potential
Taxonomic level: Application

66. *Correct answer:* **A**

Patients who are depressed will frequently manifest decreased concentration and appetite, as well as sleep disturbances, lack of interest, feelings of guilt, lack of energy, psychomotor agitation, and suicidal ideation.

> *Nursing process step:* Analysis
> *Client needs category:* Psychosocial integrity
> *Client needs subcategory:* Coping and adaptation
> *Taxonomic level:* Application

67. *Correct answer:* **B**

Respiratory tract infection is the most common acute condition experienced by elderly people.

> *Nursing process step:* Analysis
> *Client needs category:* Physiological integrity
> *Client needs subcategory:* Reduction of risk potential
> *Taxonomic level:* Knowledge

68. *Correct answer:* **B**

Progressive intellectual impairment — dementia — is gradually manifested, usually over the course of months or years. It's the most feared chronic condition among elderly people.

> *Nursing process step:* N/A
> *Client needs category:* Health promotion and maintenance
> *Client needs subcategory:* Growth and development through the life span
> *Taxonomic level:* Knowledge

69. *Correct answer:* **B**

Community health nurses focus on the prevention of disease and the promotion and maintenance of health.

> *Nursing process step:* N/A
> *Client needs category:* Health promotion and maintenance
> *Client needs subcategory:* Prevention and early detection of disease
> *Taxonomic level:* Comprehension

70. *Correct answer:* **A**

The goal of health care for elderly people in the community is to optimize their general health and functional capabilities. This will allow them to remain in their homes, which helps reduce health care costs, improves quality of life, and preserves functional status. The other responses aren't goals of health care for elderly people.

Nursing process step: N/A
Client needs category: Health promotion and maintenance
Client needs subcategory: Prevention and early detection of disease
Taxonomic level: Comprehension

71. *Correct answer:* C

Primary prevention activities are those that help reduce susceptibility to disease in people with no symptoms. Examples include prophylactic vaccines, smoking cessation classes, safety factors, dental health, exercise, and medication teaching.

Nursing process step: Planning
Client needs category: Health promotion and maintenance
Client needs subcategory: Prevention and early detection of disease
Taxonomic level: Application

72. *Correct answer:* D

Secondary prevention for elderly people includes screening for early detection, diagnosis, and treatment of disease in asymptomatic people. It has been estimated that as many as 94% of elderly people screened reveal a positive finding.

Nursing process step: Assessment
Client needs category: Health promotion and maintenance
Client needs subcategory: Prevention and early detection of disease
Taxonomic level: Application

73. *Correct answer:* D

Tertiary prevention involves maximizing independent functioning in people in whom disease is already diagnosed, including those with incontinence, early dementia, and depression.

Nursing process step: Planning
Client needs category: Psychosocial integrity
Client needs subcategory: Coping and adaptation
Taxonomic level: Application

74. *Correct answer:* B

Approximately 10% to 20% of noninstitutionalized elderly people suffer from depression.

Nursing process step: Planning
Client needs category: Health promotion and maintenance
Client needs subcategory: Prevention and early detection of disease
Taxonomic level: Knowledge

75. *Correct answer:* **D**

Approximately 20% of those over age 75 and residing in the community have some form of cognitive impairment.

> *Nursing process step:* Analysis
> *Client needs category:* Health promotion and maintenance
> *Client needs subcategory:* Prevention and early detection of disease
> *Taxonomic level:* Knowledge

76. *Correct answer:* **D**

It's best for the nurse to appear knowledgeable and comfortable with the subject of abuse, allowing the woman to share her story. Directly querying the patient regarding her injuries is the best approach. Other options allow the patient to circumvent the issue and fabricate a more acceptable cause of her injuries.

> *Nursing process step:* Implementation
> *Client needs category:* Psychosocial integrity
> *Client needs subcategory:* Psychosocial adaptation
> *Taxonomic level:* Application

77. *Correct answer:* **C**

Women who blame themselves for abuse generally have low self-esteem and suffer from depression.

> *Nursing process step:* Evaluation
> *Client needs category:* Psychosocial integrity
> *Client needs subcategory:* Coping and adaptation
> *Taxonomic level:* Comprehension

78. *Correct answer:* **D**

The most important intervention for a patient in a dangerous relationship is to provide her with resources for a safe haven, particularly if she has children. Providing couple's counseling may actually cause the abuse to worsen, and individual counseling should only be provided once the patient is safe.

> *Nursing process step:* Implementation
> *Client needs category:* Psychosocial integrity
> *Client needs subcategory:* Coping and adaptation
> *Taxonomic level:* Analysis

79. *Correct answer:* **B**

The male partner's attendance at programs that focus on his underlying feelings about women and violence are usually the most effective. These programs should be court-mandated and monitored. They don't necessarily have to be prison programs. Group therapy generally works better than individual therapy with batterers.

> *Nursing process step:* N/A
> *Client needs category:* Psychosocial integrity
> *Client needs subcategory:* Psychosocial adaptation
> *Taxonomic level:* Comprehension

80. *Correct answer:* **D**

Information regarding the patient's relationship with her spouse is important because spousal abuse is the leading cause of attempted and actual suicides in adult women.

> *Nursing process step:* Planning
> *Client needs category:* Psychosocial integrity
> *Client needs subcategory:* Psychosocial adaptation
> *Taxonomic level:* Application

81. *Correct answer:* **C**

Patients who have attempted suicide are at much higher risk for repeat attempts in the future.

> *Nursing process step:* N/A
> *Client needs category:* Psychosocial integrity
> *Client needs subcategory:* Psychosocial adaptation
> *Taxonomic level:* Comprehension

82. *Correct answer:* **A**

Although homicide is the eleventh leading cause of death among all Americans, it's the number 1 leading cause of death among young blacks of both sexes.

> *Nursing process step:* Implementation
> *Client needs category:* Psychosocial integrity
> *Client needs subcategory:* Coping and adaptation
> *Taxonomic level:* Knowledge

83. *Correct answer:* **D**

Homicide within families is usually preceded by the abuse of a family member. Thus, prevention of family homicide involves working with abusive families. The chance of eventual homicide must be kept in mind when working with these families.

> *Nursing process step:* Implementation
> *Client needs category:* Psychosocial integrity
> *Client needs subcategory:* Psychosocial adaptation
> *Taxonomic level:* Analysis

CHAPTER 5

PHYSIOLOGICAL NEEDS OF THE PATIENT

QUESTIONS

1. A patient is evaluated in the doctor's office with complaints of a nonproductive cough with low-grade fever. The nurse auscultates the patient's heart and lungs with the flat-disc diaphragm of the stethoscope. The nurse is auscultating for which of the most commonly heard sounds?

   A. High-pitched sounds
   B. Low-pitched sounds
   C. Medium-pitched sounds
   D. Moderately pitched sounds

2. A patient is admitted with a medical diagnosis of peritonitis. The basic pathophysiology of peritonitis involves:

   A. autodigestion of the pancreas.
   B. leakage of abdominal organ contents into the abdominal cavity.
   C. recurrent ulcerative and inflammatory disease of the mucosal layer of the colon.
   D. subacute and chronic inflammation that extends through all layers of the bowel.

3. A patient is admitted with a diagnosis of chronic renal failure. The nurse observes circumoral paresthesia. Circumoral paresthesia seen in patients with chronic renal failure is most commonly associated with low levels of:

   A. calcium.
   B. magnesium.
   C. phosphorus.
   D. potassium.

4. A patient is admitted to the outpatient unit for instructions in receiving peritoneal dialysis. The nurse instructs the patient to observe for which indicator of peritonitis?

   A. Bloody dialysate return
   B. Brown dialysate return
   C. Cloudy dialysate return
   D. Milky dialysate return

5. A patient is being evaluated at the clinic. On assessment the nurse observes numerous dental caries. Care of this patient should be based on the nurse's knowledge that two major types of oral problems are:

A. bleeding gums and stomatitis.
B. dental caries and halitosis.
C. dental caries and periodontal disease.
D. halitosis and gingivitis.

6. A patient is evaluated with complaints of a burning sensation in his mouth followed by the appearance of small vesicles in the oral cavity approximately 24 to 48 hours later. The nurse understands that the patient's symptoms are indicative of stomatitis, which is the result of:

A. excessive salivation.
B. irritants such as tobacco.
C. mouth breathing.
D. poor brushing and flossing.

7. When teaching personal hygiene to a female patient, the nurse understands it's important to wash the genital area from:

A. back to front.
B. front to back.
C. in a circular motion.
D. side to side.

8. A patient has just returned from the postanesthesia care unit from a hemorrhoidectomy. His postoperative orders include sitz baths every morning. The nurse provides the sitz bath with the understanding that it's used to:

A. cause swelling.
B. lower body temperature.
C. promote healing.
D. relieve tension.

9. When providing evening care for a patient, the nurse understands that the purpose of giving a patient a back massage is to:

A. promote exercise.
B. promote healing.
C. reduce infection.
D. stimulate and relax.

10. A patient is admitted to the medical-surgical unit following a total hip replacement procedure. The nurse applies an abduction pillow as ordered by the doctor. The nurse knows that abduction of a joint is described as:

A. bending the joint to form an acute angle.
B. moving the limb away from a normal position.
C. moving the limb toward a normal resting position.
D. straightening a bent joint.

11. The nurse is preparing a powdered medication for administration. Upon reconstituting a powdered medication for injection, it's necessary to use which type of solution?

A. Sterile saline
B. Sterile water
C. Spring water
D. Tap water

12. The nurse is caring for a patient with a nasogastric tube. The doctor orders a 100-ml flush through the tube every 4 hours. When it's necessary to flush a nasogastric tube, the solution to use is:

A. distilled water.
B. sterile saline.
C. sterile water.
D. tap water.

13. A child is admitted to the emergency department with a laceration to the scalp. The doctor has requested assistance while he sutures the laceration. The nurse places the patient on a papoose board with the understanding that this restraint is used with a child during a procedure:

A. that requires the patient to be motionless.
B. that causes the patient a lot of pain.
C. that causes the patient to vomit.
D. that requires sterile technique.

14. Palpation of a joint reveals crepitation. The nurse knows that crepitation found in a joint is due to:

A. bone fragments in the joint.
B. fluid in the joint.
C. infection in the joint.
D. normal functioning.

15. Chvostek's sign is associated with which electrolyte imbalance?

A. Hypocalcemia
B. Hypokalemia
C. Hyponatremia
D. Hypophosphatemia

16. A patient is admitted to the medical-surgical unit following surgery. Four days after surgery, the patient spikes a 102° F (38.9° C) oral temperature and exhibits a wet, productive cough. The nurse assesses the patient with the understanding that an infection that is acquired during hospitalization is known as:

    A. a community-acquired infection.
    B. an iatrogenic infection.
    C. a nosocomial infection.
    D. an opportunistic infection.

17. When caring for a patient who is at risk for falling, it's important to perform which of the following tasks?

    A. Keep all objects away from the patient so he doesn't fall on them.
    B. Keep the bed in the high position so the patient won't want to get out of it.
    C. Raise the side rails on beds and stretchers when appropriate.
    D. Use throw rugs on the area around the bed so the patient's feet will be warm when standing.

18. The use of isolation in the hospital is intended to:

    A. discourage a patient with an infection from ambulating.
    B. keep an infection from becoming endemic.
    C. maintain a sterile environment.
    D. prevent the further spread of an infection to others.

19. The nurse is caring for a patient who develops sudden weakness on one side of his body and facial droop. This would most likely be associated with the patient having a:

    A. cerebrovascular accident.
    B. muscle spasm.
    C. myocardial infarction.
    D. pulmonary embolism.

20. A patient is admitted to the emergency department with complaints of chest pain and shortness of breath. The nurse's assessment reveals jugular venous distention. The nurse knows that when a patient has jugular venous distention it's typically due to:

    A. a neck tumor.
    B. an electrolyte imbalance.
    C. dehydration.
    D. fluid overload.

21. The nurse's assessment reveals that the patient has exophthalmos. The nurse knows that exophthalmos is most likely associated with what disorder?

    A. Hypernatremia
    B. Hyperthyroidism
    C. Hyponatremia
    D. Hypothyroidism

22. When performing an extraocular movement examination, the patient exhibits nystagmus. This can best be described as:

    A. crossing of the eyes.
    B. excessive tearing of the eye.
    C. flickering of the eye lid.
    D. jerking movements of the eyeball.

23. A nurse is performing a breast examination. She observes a nipple discharge in the patient, who isn't lactating. The nurse knows that the patient:

    A. is sexually aroused.
    B. has breast retraction.
    C. may be pregnant.
    D. may have a breast malignancy.

24. The nurse is assessing a patient who has a diagnosis of scoliosis. The nurse knows that scoliosis can be described as:

    A. an anterior curvature of the spine.
    B. a humpback.
    C. a lateral curvature of the spine.
    D. a swayback.

25. The nurse is examining a patient involved in a car accident. The patient complains of shortness of breath. The tissue of the patient's face, neck, and torso gives a crackling sensation when palpated. The nurse suspects subcutaneous emphysema. Subcutaneous emphysema is due to:

    A. air under the skin.
    B. fluid in the lungs.
    C. long-term smoking.
    D. using bronchodilators for several years.

26. An indication for an Allen's test to be performed is that the patient:

    A. has peripheral neuropathy.
    B. has ulnar nerve damage.
    C. needs a glucose test.
    D. needs an arterial line placed.

27. The nurse is caring for a patient with complaints of headache and neck pain. Assessment reveals painful flexion of the neck to the chest. The nurse understands that nuchal rigidity is associated with:

    A. brain tumor.
    B. cerebrovascular accident.
    C. meningitis.
    D. subdural hematoma.

28. If the patient is experiencing pain, numbness, and tingling of the hand, the patient may have:

    A. carpal tunnel syndrome.
    B. peripheral neuropathy.
    C. Raynaud's disease.
    D. tendonitis.

29. A patient with rheumatoid arthritis can be expected to have which of the following signs?

    A. Bouchard's nodes
    B. Heberden's nodes
    C. Spindle-shaped fingers
    D. Swollen hands and feet

30. Sharp pain in the heel when walking can be due to:

    A. Achilles tendonitis.
    B. arthritis.
    C. joint deformity.
    D. joint infection.

31. Trousseau's sign is associated with which electrolyte imbalance?

    A. Hyperkalemia
    B. Hypercalcemia
    C. Hypocalcemia
    D. Hyponatremia

32. A shuffling gait is typically associated with the patient who has:

    A. multiple sclerosis.
    B. Parkinson's disease.
    C. residual weakness from a stroke.
    D. rheumatoid arthritis.

33. The patient with an acoustic nerve injury may have a positive Romberg's sign. A positive Romberg's sign is the inability to:

    A. hear a whispered voice.
    B. hear unless the listener is looking at the speaker.
    C. maintain an upright position with eyes closed.
    D. maintain an upright position with eyes open.

34. When performing a neurological exam, a plantar reflex can be performed. If the patient exhibits a positive Babinski reflex, this reaction is:

    A. abnormal and indicates cortical suppression.
    B. abnormal and indicates spinal injury.
    C. normal.
    D. normal if the patient is over age 2.

35. When a patient experiences a loss of vibratory sense on examination, this indicates:

    A. injury to the cranial nerves.
    B. injury to the peripheral nerves.
    C. intact cranial nerves.
    D. intact peripheral nerves.

36. Taking verbal orders over the phone can be a problem because:

    A. there's no written record of the order.
    B. the order isn't valid if it's not written down.
    C. the patient doesn't get charged for medications that aren't written down.
    D. the pharmacy won't provide the medication if the prescription isn't written by a doctor

SITUATION: *Maggie Simmons is a 24-year-old teacher admitted to the hospital with acute appendicitis. She's now 1-day postoperative appendectomy and has an abdominal surgical wound.*

    Questions 37 to 39 relate to this situation.

37. Based on the nurse's knowledge of surgical wounds, simple surgical incisions heal by _____ intention.

    A. primary
    B. quarternary
    C. secondary
    D. tertiary

38. The nurse documents that the wound edges are approximated. When the edges of an incision are said to be approximated, this means the edges are:

    A. brought together, usually by sutures, tape, or staples.
    B. erythematous and swollen.
    C. gaping and draining.
    D. necrotic and draining.

39. The nurse assesses Miss Simmons's wound and notes that scar tissue has begun to form. The formation of scar tissue in wound healing is secondary to healing by:

    A. application of skin grafts.
    B. collagen synthesis.
    C. epithelization.
    D. phagocytosis.

SITUATION: *Fanny McAlister is an 85-year-old woman who has been bedridden for several months. The nurse is providing wound care for a stage III decubitus ulcer.*

    Questions 40 to 43 relate to this situation.

40. Assessment of the wound reveals the formation of eschar and signs of infection. The formation of eschar occurs when a wound is:

    A. healing by primary intention.
    B. necrotic and infected.
    C. sutured closed.
    D. unable to heal by epithelization.

41. Which nutrient is most needed for wound healing?

    A. Carbohydrates
    B. Fat
    C. Protein
    D. Vitamins

42. The doctor has scheduled Mrs. McAlister for a surgical wound debridement. The nurse knows that debridement means:

    A. packing the wound with gauze.
    B. removing necrotic tissue.
    C. reparation of the wound.
    D. suturing the wound edges.

43. When assessing Mrs. McAlister's wound for signs of infection, the nurse should look for the presence of which of the following signs?

    A. Granulation tissue
    B. Pink tissue
    C. Purulent drainage
    D. Well-approximated edges

SITUATION: *Herman Fleck is a 65-year-old man discharged from the hospital to home care for complete bed rest. The nurse's plan of care includes passive range-of-motion exercises. This plan of care is based on the nurse's knowledge of the effects of prolonged bed rest.*

Questions 44 to 51 relate to this situation.

44. A way to prevent foot drop in a patient requiring bed rest is to:

    A. Gatch the knee of the bed.
    B. place a footboard firmly against the feet.
    C. place a roll under the ankles.
    D. place a trochanter roll under the thighs.

45. The effects of prolonged bed rest on the musculoskeletal system include muscle:

    A. agility.
    B. build-up.
    C. weakness.
    D. strengthening.

46. The nurse understands that pressure sores are most likely to develop over which part of the body?

    A. Abdomen
    B. Area around the ears
    C. Chest
    D. Bony prominences

47. Passive range-of-motion exercises are performed to:

    A. maintain joint immobility.
    B. overstimulate circulation.
    C. prevent strengthening of muscles.
    D. restore loss of joint function.

48. When performing passive range-of-motion exercises, the limb should always be moved:

    A. to the point of fatigue.
    B. to the point of meeting resistance.
    C. to the point of pain.
    D. through the point of pain.

49. Mr. Fleck has had his activity level increased, and he tells the nurse that upon rising he becomes light-headed and dizzy. The nurse understands that these symptoms are characteristic of:

    A. orthostatic hypertension.
    B. orthostatic hypotension.
    C. orthostatic stasis.
    D. venous stasis.

50. The nurse's plan of care for Mr. Fleck will also include nursing interventions to prevent deep vein thrombosis (DVT) and common disorders associated with immobility. The nurse knows that DVT can occur when a patient experiences extended bed rest. This is due to the development of:

    A. increased anticoagulation.
    B. increased blood viscosity.
    C. increased coagulability.
    D. stasis ulcers.

51. A common GI disorder associated with immobility is:

    A. constipation.
    B. diarrhea.
    C. gastric reflux.
    D. increased acid production.

SITUATION: *Richard Bradley is a 44-year-old jeweler admitted to the emergency department complaining of abdominal pain, nausea, and vomiting.*

    Questions 52 to 58 relate to this situation.

52. When performing an abdominal examination, Mr. Bradley should be in a supine position with the head of the bed at what position?

    A. 0 degrees
    B. 30 degrees
    C. 45 degrees
    D. 90 degrees

53. The nurse knows that the order of physical assessment of the abdomen is:

    A. auscultation, observation, percussion, and palpation.
    B. observation, auscultation, percussion, and palpation.
    C. observation, percussion, palpation, and auscultation.
    D. percussion, observation, auscultation, and palpation.

54. During the admission assessment, the nurse percusses over muscle. The nurse understands that when percussing over a dense structure such as a muscle, the sound heard would be:

    A. dullness.
    B. flatness.
    C. resonance.
    D. tympany.

55. On palpation of Mr. Bradley's abdomen, the nurse observes guarding. When a patient exhibits guarding on abdominal assessment, the patient is:

    A. consciously tightening abdominal muscles.
    B. experiencing pain in the right lower quadrant when the left lower quadrant is palpated.
    C. in extreme pain on palpation.
    D. in pain when palpation is quickly released.

56. Observation of Mr. Bradley's skin reveals generalized jaundice. The nurse knows that jaundice is most commonly associated with:

    A. acute renal failure.
    B. diabetes mellitus.
    C. heart failure.
    D. liver failure.

57. Mr. Bradley is being evaluated for possible cholelithiasis. The nurse observes icteric sclera. The nurse knows that icteric sclera can be associated with:

    A. abdominal distention.
    B. jaundice.
    C. low urine output.
    D. peripheral edema.

58. The most accurate measurement of fluid status in Mr. Bradley is:

    A. anthropometric measurements.
    B. blood pressure.
    C. body temperature.
    D. body weight.

SITUATION: *Steven Burns is a 68-year-old baker admitted to the medical unit with a diagnosis of cerebral vascular accident (stroke). Assessment reveals that the patient is unresponsive to verbal or painful stimuli.*

Questions 59 to 61 relate to this situation.

59. The nurse understands that the correct position to perform mouth care on a comatose patient is:

    A. side-lying.
    B. semi-Fowler's.
    C. prone.
    D. supine.

60. The nurse is providing morning care to Mr. Burns. The nurse understands that the direction of washing the eye should be from:

    A. inner canthus to outer canthus.
    B. lower canthus to upper canthus.
    C. outer canthus to inner canthus.
    D. upper canthus to lower canthus.

61. The nurse is aware that it's important to moisturize the eyes of a comatose patient and to keep them closed to prevent:

    A. conjunctival hemorrhage.
    B. corneal abrasion.
    C. pupil dilation.
    D. retina detachment.

SITUATION: *The nurse has recently taken a job in a local hospital in the operating room. A surgical scrub nurse understands that principles of asepsis and sterile technique must be implemented to prevent the spread of infection.*

Questions 62 to 67 relate to this situation.

62. The nurse understands that surgical asepsis means:

    A. free of all pathogenic organisms.
    B. free of most pathogenic organisms.
    C. having pathogenic organisms.
    D. removing pathogenic organisms.

63. Sterilization is the process by which an object:

    A. becomes free from all pathogenic organisms.

    B. is boiled in water to remove pathogenic organisms.

    C. is steamed in an effort to remove all pathogenic organisms.

    D. is washed with soap to remove pathogenic organisms.

64. The nurse has a sterile field in front of her and needs to reach something on the other side of the sterile field. To maintain sterile technique, the nurse should:

    A. move the sterile field away from her and reach the object.

    B. reach across the sterile field.

    C. walk around the sterile field, keeping herself facing the sterile field.

    D. walk around the sterile field with her back to the sterile field.

65. The nurse is aware that cold sterilization or soaking is used for:

    A. all glass objects.

    B. all metal objects.

    C. all wooden objects.

    D. rubber and plastic objects.

66. The nurse is aware that dry-heat sterilization, or baking, is used for objects that:

    A. are made of glass.

    B. are made of metal.

    C. are made of rubber.

    D. have sharp edges.

67. When preparing equipment for surgical procedures, the nurse knows that sterile objects remain sterile only as long as they're in a package with an impervious cover. An impervious cover is one that:

    A. is open.

    B. is waterproof.

    C. has a torn cover.

    D. has been tampered with.

SITUATION: *Pauline Hack is a 40-year-old primigravida. She is in the clinic for an amniocentesis. The nurse is preparing to assist the doctor while he performs the amniocentesis.*

    Questions 68 to 70 relate to this situation.

68. When preparing for surgical or sterile procedures, the nurse knows that a sterile field is a work surface area prepared with:

    A. clean cloths.
    B. clean towels.
    C. sterile drapes.
    D. sterile packages.

69. The proper way to open an envelope-wrapped sterile package after removing the outer package or tape is to open the first portion of the wrapper:

    A. away from the body.
    B. to the left of the body.
    C. to the right of the body.
    D. toward the body.

70. If required to place a sterile item on a sterile field, the nurse should:

    A. hand the sterile package to a sterile nurse and have her place it on the sterile field.
    B. open the sterile package and drop the item on the sterile field.
    C. open the sterile package and place the item on the sterile field.
    D. open the sterile package and take the item out with a clean, gloved hand.

SITUATION: *Sharon Rossanelli's doctor has ordered an insertion of a urinary catheter as part of her preoperative orders.*

    Questions 71 to 77 relate to this situation.

71. When preparing a sterile field, the best way for the nurse to remain sterile is to:

    A. put on sterile gloves, open all the sterile containers with one gloved hand, and prepare the sterile field with the other gloved hand.
    B. put on sterile gloves, prepare the sterile field, and open all the sterile containers.
    C. open all the sterile containers, put on sterile gloves, and prepare the sterile field.
    D. open all sterile containers, prepare the sterile field, and put on sterile gloves.

72. When putting on sterile gloves, the nurse should open the outer package and ensure that the glove cuffs are facing her. She should then open the outer wrap. The next step the nurse should perform is to:

A. pick up both gloves by the cuff and determine which is right and which is left.
B. pick up the first glove by holding on to the wrap and shaking it loose.
C. pick up the first glove by sliding her hand under the cuff.
D. pick up the first glove by the folded back cuff.

73. When the nurse places her hand into a sterile glove, it's best for her to have the fingers of the glove pointing:

A. down to the floor.
B. palm down.
C. palm up.
D. up to the sky.

74. After sterile gloves are put on, it's important to keep one's hands:

A. down and below the waist.
B. folded and above the waist.
C. folded and below the waist.
D. in sight and above the waist.

75. When pouring sterile liquids, the proper technique is to open the bottle and then:

A. lean the bottle rim on the sterile container and pour the liquid into the container.
B. place the cap face down and pour the liquid into the sterile container.
C. place the cap face down, lean the bottle rim on the sterile container, and pour the liquid into the container.
D. place the cap face up and pour the liquid above the sterile container.

76. When preparing to clean the area of the urinary meatus before inserting a catheter, the nondominant hand is used to separate the labia and expose the meatus. The nondominant hand:

A. may touch sterile items once the cleaning is finished.
B. may touch sterile items only if absolutely necessary.
C. may touch the urinary catheter upon inserting it.
D. must not touch anything sterile.

77. When obtaining a urine sample from a patient with an indwelling urinary catheter, the urine should be taken:

    A. from the balloon port of the catheter.
    B. from the collection bag.
    C. from the collection port of the catheter with a needle.
    D. only when first inserting the catheter.

SITUATION: *Mark Honer is a 25-year-old male diagnosed with delusional disorder. He has been extremely combative, and the doctor has written an order for wrist restraints.*

Questions 78 to 80 relate to this situation.

78. When applying wrist restraints to a combative patient, the restraints should be fastened to the:

    A. bed frame on the opposite side of the bed.
    B. bed frame on the same side of the bed.
    C. side rail on the opposite side of the bed.
    D. side rail on the same side of the bed.

79. When a patient is wearing wrist or ankle restraints, the restraints should be released every:

    A. hour.
    B. 2 hours.
    C. 4 hours.
    D. 8 hours.

80. If Mr. Honer requires further restraint, the nurse knows that a chest or vest restraint is used to prevent the patient from:

    A. getting out of bed.
    B. moving a part of his body.
    C. pulling out tubes.
    D. rolling over in bed.

ANSWER SHEET

	A B C D		A B C D		A B C D		A B C D
1	○ ○ ○ ○	21	○ ○ ○ ○	41	○ ○ ○ ○	61	○ ○ ○ ○
2	○ ○ ○ ○	22	○ ○ ○ ○	42	○ ○ ○ ○	62	○ ○ ○ ○
3	○ ○ ○ ○	23	○ ○ ○ ○	43	○ ○ ○ ○	63	○ ○ ○ ○
4	○ ○ ○ ○	24	○ ○ ○ ○	44	○ ○ ○ ○	64	○ ○ ○ ○
5	○ ○ ○ ○	25	○ ○ ○ ○	45	○ ○ ○ ○	65	○ ○ ○ ○
6	○ ○ ○ ○	26	○ ○ ○ ○	46	○ ○ ○ ○	66	○ ○ ○ ○
7	○ ○ ○ ○	27	○ ○ ○ ○	47	○ ○ ○ ○	67	○ ○ ○ ○
8	○ ○ ○ ○	28	○ ○ ○ ○	48	○ ○ ○ ○	68	○ ○ ○ ○
9	○ ○ ○ ○	29	○ ○ ○ ○	49	○ ○ ○ ○	69	○ ○ ○ ○
10	○ ○ ○ ○	30	○ ○ ○ ○	50	○ ○ ○ ○	70	○ ○ ○ ○
11	○ ○ ○ ○	31	○ ○ ○ ○	51	○ ○ ○ ○	71	○ ○ ○ ○
12	○ ○ ○ ○	32	○ ○ ○ ○	52	○ ○ ○ ○	72	○ ○ ○ ○
13	○ ○ ○ ○	33	○ ○ ○ ○	53	○ ○ ○ ○	73	○ ○ ○ ○
14	○ ○ ○ ○	34	○ ○ ○ ○	54	○ ○ ○ ○	74	○ ○ ○ ○
15	○ ○ ○ ○	35	○ ○ ○ ○	55	○ ○ ○ ○	75	○ ○ ○ ○
16	○ ○ ○ ○	36	○ ○ ○ ○	56	○ ○ ○ ○	76	○ ○ ○ ○
17	○ ○ ○ ○	37	○ ○ ○ ○	57	○ ○ ○ ○	77	○ ○ ○ ○
18	○ ○ ○ ○	38	○ ○ ○ ○	58	○ ○ ○ ○	78	○ ○ ○ ○
19	○ ○ ○ ○	39	○ ○ ○ ○	59	○ ○ ○ ○	79	○ ○ ○ ○
20	○ ○ ○ ○	40	○ ○ ○ ○	60	○ ○ ○ ○	80	○ ○ ○ ○

ANSWERS AND RATIONALES

1. *Correct answer:* **A**

The flat-disc diaphragm is best used for transmission of high-pitched (bronchial) sounds. The bell-shaped diaphragm is used to transmit low-pitched sounds, such as heart sounds.

> *Nursing process step:* Assessment
> *Client needs category:* Physiological integrity
> *Client needs subcategory:* Physiological adaptation
> *Taxonomic level:* Knowledge

2. *Correct answer:* **B**

Peritonitis results from the leakage of abdominal organ contents into the abdominal cavity, usually resulting from inflammation, infection, trauma, or perforation. Autodigestion of the pancreas, usually by its own enzymes (principally trypsin), is characteristic of pancreatitis. A recurrent ulcerative and inflammatory disease of the mucosal layer of the colon is characteristic of the pathophysiology of ulcerative colitis. A subacute and chronic inflammation that extends through all layers of the bowel is associated with regional enteritis.

> *Nursing process step:* Planning
> *Client needs category:* Physiological integrity
> *Client needs subcategory:* Physiological adaptation
> *Taxonomic level:* Application

3. *Correct answer:* **A**

Low levels of calcium are common in chronic renal failure due to high phosphorus levels. The word *circumoral* refers to encircling the mouth. Numbness and tingling in extremities around the mouth as well as dizziness are associated with low calcium levels. Muscle weakness and leg cramps are associated with low potassium levels; decreased muscle strength and muscle pain are characteristic of low phosphorus levels. Clinical signs associated with low magnesium levels are neuromuscular irritability with tremors and increased reflexes.

> *Nursing process step:* Assessment
> *Client needs category:* Physiological integrity
> *Client needs subcategory:* Physiological adaptation
> *Taxonomic level:* Comprehension

4. *Correct answer:* **C**

Peritonitis and abdominal infection cause the dialysate fluid to turn cloudy. Bloody drainage may be observed occasionally with young menstruating women, catheter displacement, or minor trauma, or during the first few exchanges.

> *Nursing process step:* Implementation
> *Client needs category:* Physiological integrity
> *Client needs subcategory:* Reduction of risk potential
> *Taxonomic level:* Comprehension

5. *Correct answer:* **C**

Dental caries are responsible for most of the tooth loss before age 35, and periodontal disease is the principal cause of tooth loss after age 35. Halitosis, or bad breath, is usually associated with poor oral hygiene. Bleeding gums and stomatitis are observed in infectious processes and immunosuppressed patients.

> *Nursing process step:* Planning
> *Client needs category:* Physiological Integrity
> *Client needs subcategory:* Physiological adaptation
> *Taxonomic level:* Comprehensive

6. *Correct answer:* **B**

Stomatitis is an inflammatory process of the oral mucosa and can result from irritants—such as tobacco or medications—which result in immunosuppression. Poor brushing and flossing result in dental caries and periodontal disease.

> *Nursing process step:* Assessment
> *Client needs category:* Physiological integrity
> *Client needs subcategory:* Physiological adaptation
> *Taxonomic level:* Comprehensive

7. *Correct answer:* **B**

The urethral orifice is considered the cleanest area, and the anal orifice is considered the least clean area, so it's best to wash from clean to dirty, or front to back. Cleansing the perineal area from side to side, back to front, or in a circular motion would promote the transmission of microorganisms from areas of contamination to the urethral orifice.

> *Nursing process step:* Implementation
> *Client needs category:* Physiological integrity
> *Client needs subcategory:* Reduction of risk potential
> *Taxonomic level:* Comprehension

8. *Correct answer:* **C**

Sitz baths provides moist heat to the perineal and anal area to clean, promote healing and drainage, and reduce soreness to the area. Sitz baths help healing with cleaning action and promotion of circulation, thereby reducing swelling. Sitz baths usually have no therapeutic value in lowering body temperature. Although relief of tension can occur, this effect is secondary to the promotion of healing.

> *Nursing process step:* Implementation
> *Client needs category:* Health promotion and maintenance
> *Client needs subcategory:* Growth and development through the life span
> *Taxonomic level:* Comprehensive

9. *Correct answer:* **D**

The major purpose of a massage is to promote increased circulation, relieve muscle tension, promote physical and mental relaxation, improve muscle and skin functioning, relieve insomnia, and provide relief from pain. Back massages usually don't have therapeutic value in reducing infection or promoting exercise. Promotion of healing is secondary to the improvement of muscle and skin functioning and the relief of insomnia.

> *Nursing process step:* Implementation
> *Client needs category:* Physiological integrity
> *Client needs subcategory:* Basic care and comfort
> *Taxonomic level:* Comprehensive

10. *Correct answer:* **B**

Abduction is when an arm, leg, or finger is moved away from the normal resting position.

> *Nursing process step:* Implementation
> *Client needs category:* Physiological integrity
> *Client needs subcategory:* Reduction of risk potential
> *Taxonomic level:* Knowledge

11. *Correct answer:* **B**

The best solution for reconstituting a powdered medicine is sterile water. Medications for administration must remain sterile. Tap water and spring water are considered nonsterile. Sterile sodium chloride (saline) isn't considered the best solution for mixing medication.

> *Nursing process step:* Implementation
> *Client needs category:* Physiological integrity
> *Client needs subcategory:* Pharmacological and parenteral therapies
> *Taxonomic level:* Application

12. *Correct answer:* **D**

The stomach isn't considered sterile, so using tap water is acceptable, unless some underlying condition exists that prohibits its use. Flushing the tube with sterile saline may cause GI or electrolyte problems.

Nursing process step: Implementation
Client needs category: Physiological integrity
Client needs subcategory: Reduction of risk potential
Taxonomic level: Application

13. *Correct answer:* **A**

A papoose board is used with a child for a procedure requiring the child to be motionless, such as venipuncture or suturing. The papoose board restrains the child in a supine position. It's contraindicated for vomiting patients to minimize the risk of aspiration. Papoose boards may be utilized in painful procedures or in procedures that require sterile technique, but they aren't necessary for those procedures.

Nursing process step: Implementation
Client needs category: Safe, effective care environment
Client needs subcategory: Safety and infection control
Taxonomic level: Comprehensive

14. *Correct answer:* **A**

The presence of crepitation in a joint means that there are loose bodies in the joint, typically bone fragments from the friction of the opposing bones rubbing one another. Fluid in the joint may result in painful or difficult movement, but it doesn't cause a crepitation. Infection in the joint may result in swelling and painful or difficult movement, but it doesn't result in a crepitation. Crepitation in a joint is an abnormal finding and requires further evaluation.

Nursing process step: Assessment
Client needs category: Physiological integrity
Client needs subcategory: Physiological adaptation
Taxonomic level: Knowledge

15. *Correct answer:* **A**

Chvostek's sign is a spasm of the facial muscles elicited by tapping the facial nerve and is associated with hypocalcemia. Clinical signs of hypokalemia are muscle weakness, leg cramps, fatigue, nausea, and vomiting. Muscle cramps, anorexia, nausea, and vomiting are clinical signs of hyponatremia. Clinical manifestations associated with hypophosphatemia include muscle pain, confusion, seizures, and coma.

Nursing process step: Assessment
Client needs category: Physiological integrity
Client needs subcategory: Physiological adaptation
Taxonomic level: Knowledge

16. *Correct answer:* **C**

Nosocomial, or hospital-acquired, infections are infections acquired during hospitalization for which the patient isn't being primarily treated. Community-acquired or opportunistic infections may not be acquired during hospitalization. An iatrogenic infection is caused by the doctor or by medical therapy, and an opportunistic infection affects a compromised host.

Nursing process step: Assessment
Client needs category: Safe, effective care environment
Client needs subcategory: Safety and infection control
Taxonomic level: Knowledge

17. *Correct answer:* **C**

The use of side rails on beds and stretchers will help remind the patient that he should not get out of bed without assistance. Items the patient may want to use should be kept close to him so he can reach them. It's also important to keep the bed in the low position; in the event the patient does try to get out of bed, he won't fall as far. Throw rugs should be avoided because they tend to slip out from under the patient upon standing.

Nursing process step: Implementation
Client needs category: Safe, effective care environment
Client needs subcategory: Safety and infection control
Taxonomic level: Application

18. *Correct answer:* **D**

Isolation protocols are used in the hospital setting to contain an organism to one area and reduce the spread of infection to others. Isolation protocols may not prevent an infection from becoming endemic, nor will they maintain a sterile environment. Patients may ambulate while on isolation protocols, but they must be required to follow the precautions specified by the specific type of isolation in effect.

Nursing process step: Implementation
Client needs category: Safe, effective care environment
Client needs subcategory: Safety and infection control
Taxonomic level: Comprehensive

19. *Correct answer:* **A**

A patient experiencing a cerebrovascular accident typically exhibits weakness on one side of the body as well as facial droop. Depending on the area of the brain involved, there may be other symptoms. Muscle spasms usually will result in temporary pain in the affected site. A patient experiencing a myocardial infarction will usually exhibit chest pain and shortness of breath, but he shouldn't be affected with weakness on one side of the body. A patient diagnosed with a pulmonary embolism will usually complain of chest pain and shortness of breath; one-sided weakness and facial droop are usually not concerns for these patients.

> *Nursing process step:* Assessment
> *Client needs category:* Physiological integrity
> *Client needs subcategory:* Physiological adaptation
> *Taxonomic level:* Comprehension

20. *Correct answer:* **D**

Fluid overload causes the volume of blood within the vascular system to increase. This increase causes the veins to distend and can be seen most obviously in the neck veins. An electrolyte imbalance may result in fluid overload, but it doesn't directly contribute to jugular venous distention. Dehydration doesn't cause jugular venous distention. A neck tumor may result in swelling to the neck, but it won't cause jugular venous distention.

> *Nursing process step:* Assessment
> *Client needs category:* Physiological integrity
> *Client needs subcategory:* Physiological adaptation
> *Taxonomic level:* Comprehension

21. *Correct answer:* **B**

Hyperthyroidism is a condition where the thyroid gland overproduces thyroid hormones. This overproduction can result in such physical signs as tachycardia, excessive diaphoresis, exophthalmos (protrusion of the eyeball), and an enlarged thyroid gland. Clinical signs associated with hyponatremia include muscle cramps, anorexia, nausea, and vomiting. Extreme thirst; dry, sticky mucous membranes; and a dry, swollen tongue are signs associated with hypernatremia. Clinical manifestations of hair loss, brittle nails, and dry skin are associated with hypothyroidism.

> *Nursing process step:* Assessment
> *Client needs category:* Physiological integrity
> *Client needs subcategory:* Physiological adaptation
> *Taxonomic level:* Comprehension

22. *Correct answer:* **D**

Nystagmus is a jerking movement of the eyeball and is associated with a disorder of cranial nerves II, III, and VI. Crossing of the eyes is known as strabismus.

> *Nursing process step:* Assessment
> *Client needs category:* Physiological integrity
> *Client needs subcategory:* Physiological adaptation
> *Taxonomic level:* Comprehension

23. *Correct answer:* **D**

A nipple discharge is considered an abnormal finding in a nonlactating female patient and may be related to many causes. Certain medications, frequent breast stimulation, and movement of breast tissue during aerobic exercise may be contributing factors; however, the patient may have a breast malignancy and should be referred for further evaluation.

> *Nursing process step:* Assessment
> *Client needs category:* Health promotion and maintenance
> *Client needs subcategory:* Prevention and early detection of disease
> *Taxonomic level:* Comprehension

24. *Correct answer:* **C**

Scoliosis is a lateral curvature of the thoracic spine and can be hereditary. A humpback results from the gradual collapse of a vertebrae and is known as kyphosis. A swayback or the exaggeration of the lumbar spine curve is referred to as lordosis.

> *Nursing process step:* Assessment
> *Client needs category:* Physiological integrity
> *Client needs subcategory:* Physiological adaptation
> *Taxonomic level:* Knowledge

25. *Correct answer:* **A**

A patient with subcutaneous emphysema has air under his skin. This is typically due to a pneumothorax, which has allowed air to leak from the pleural space into the subcutaneous tissue, typically of the thorax. Long-term smoking and using bronchodilators for several years are related to chronic obstructive pulmonary disease. Having fluid in the lungs is related to heart failure.

> *Nursing process step:* Assessment
> *Client needs category:* Physiological integrity
> *Client needs subcategory:* Physiological adaptation
> *Taxonomic level:* Comprehension

26. *Correct answer:* **D**

An Allen's test is performed to test the patency of the ulnar artery in case the radial arterial line becomes clotted. By ensuring ulnar blood flow, the test ensures that there will be continued blood flow to the hand if clotting occurs. The Allen's test isn't used to assess nerve damage. Peripheral neuropathy is a disorder affecting the peripheral motor, sensory, or autonomic nerves that can result in loss of sensation, muscle weakness, diminished reflexes, and pain. A glucose test is a laboratory diagnostic examination performed on a venous blood sample and is used to assess for diabetes mellitus.

> *Nursing process step:* Assessment
> *Client needs category:* Physiological integrity
> *Client needs subcategory:* Reduction of risk potential
> *Taxonomic level:* Comprehension

27. *Correct answer:* **C**

A patient with meningitis will exhibit signs that include photophobia and nuchal rigidity, which is pain on flexion of the chin to the chest. Cerebrovascular accident is usually associated with hemiparesis. Brain tumors can produce sensory and motor abnormalities, visual alterations, seizures, and symptoms related to increased intracranial pressure. Clinical manifestations of subdural hematoma can range from headache and seizures to coma, depending on whether the condition is acute or chronic.

> *Nursing process step:* Assessment
> *Client needs category:* Physiological integrity
> *Client needs subcategory:* Physiological adaptation
> *Taxonomic level:* Comprehension

28. *Correct answer:* **A**

Carpal tunnel syndrome is due to the compression of the carpal bones on the nerves of the hand causing pain, numbness, and tingling of the hand. Peripheral neuropathy is a disorder affecting the peripheral motor, sensory, or autonomic nerves that can result in loss of sensation, muscle weakness, diminished reflexes, and pain. Raynaud's disease is a form of intermittent arteriolar vasoconstriction that results in coldness, pain, and pallor of the fingertips, toes, or tip of the nose. Tendonitis—the inflammation of a tendon—is due to overuse of the tendon.

> *Nursing process step:* Assessment
> *Client needs category:* Physiological integrity
> *Client needs subcategory:* Basic care and comfort
> *Taxonomic level:* Comprehension

29. *Correct answer:* **C**

A patient with rheumatoid arthritis affecting the hands will have spindle-shaped fingers due to swelling of the interphalangeal joints. Swollen hands and feet can be related to cardiovascular disorders. Bouchard's and Heberden's nodes are seen in osteoarthritis.

> *Nursing process step:* Assessment
> *Client needs category:* Physiological integrity
> *Client needs subcategory:* Physiological adaptation
> *Taxonomic level:* Comprehension

30. *Correct answer:* **A**

Achilles tendonitis is a common disorder of overweight patients or patients with poorly fitting shoes. Clinical manifestations associated with arthritis are pain, stiffness, and functional impairment. Clinical manifestations associated with joint deformity and infection are limited motion, stiffness, and swelling.

> *Nursing process step:* Assessment
> *Client needs category:* Physiological integrity
> *Client needs subcategory:* Basic care and comfort
> *Taxonomic level:* Comprehension

31. *Correct answer:* **C**

Trousseau's sign is a carpal pedal spasm elicited when a blood pressure cuff is inflated on the arm of a patient with hypocalcemia. Clinical signs of hyperkalemia include irritability, apathy, confusion, and cardiac arrhythmias. Symptoms of hypercalcemia include lethargy and weakness. Hyponatremic patients may present with muscle cramps, anorexia, nausea, and vomiting.

> *Nursing process step:* Assessment
> *Client needs category:* Physiological integrity
> *Client needs subcategory:* Physiological adaptation
> *Taxonomic level:* Knowledge

32. *Correct answer:* **B**

A shuffling gait from the musculoskeletal rigidity of the patient with Parkinson's disease is common. Patients experiencing a stroke usually exhibit loss of voluntary control over motor movements associated with generalized weakness; a shuffling gait is usually not observed in stroke patients. The most common clinical symptoms of multiple sclerosis are fatigue, numbness, difficulty in coordination, and loss of balance. Joint pain and swelling and joint stiffness are common in patients with rheumatoid arthritis.

Nursing process step: Assessment
Client needs category: Physiological integrity
Client needs subcategory: Physiological adaptation
Taxonomic level: Comprehension

33. *Correct answer:* **C**

A positive Romberg's sign means the patient is unable to maintain an upright position with feet together and eyes closed due to a vestibular disorder or an injury to the acoustic nerve. Auditory acuity is measured by the whisper test.

Nursing process step: Assessment
Client needs category: Physiological integrity
Client needs subcategory: Physiological adaptation
Taxonomic level: Comprehension

34. *Correct answer:* **A**

Dorsiflexion of the big toe when the sole of the foot is stimulated indicates suppressed cortical control and is a pathologic response in a person over age 2. A positive Babinski reflex is more indicative of a central nervous system disease affecting the corticospinal tract.

Nursing process step: Assessment
Client needs category: Health promotion and maintenance
Client needs subcategory: Prevention and early detection of disease
Taxonomic level: Comprehensive

35. *Correct answer:* **B**

Appropriate perception of vibration indicates intact dorsal column tracts and peripheral nerves. If there is a loss of vibratory sense, an injury to the peripheral nerves is probable.

Nursing process step: Assessment
Client needs category: Physiological integrity
Client needs subcategory: Physiological adaptation
Taxonomic level: Comprehension

36. *Correct answer:* **A**

Taking verbal orders over the phone can be a problem because there could be an error in interpretation of the order. Because the order wasn't written down, there is no written record to check. It's important to avoid verbal orders whenever possible or to get the order in writing as soon as possible after the order is given. Verbal orders are valid, but most institutions require that they be signed by a doctor within 24 hours. Charging for medications should not affect the safety of administration procedures. Most agency pharmacies respect verbal medication orders and will provide those medications.

Nursing process step: Implementation
Client needs category: Safe, effective care environment
Client needs subcategory: Safety and infection control
Taxonomic level: Knowledge

37. *Correct answer:* **A**

Primary intention healing occurs where the tissue surfaces have been approximated and there has been minimal or no loss of tissue. A surgical incision has minimal tissue loss and heals through the process of collagen synthesis, also called primary intention. In secondary intention, the edges cannot be approximated and considerable tissue loss occurs. An example of healing by secondary intention would be a decubitus ulcer. Tertiary healing occurs when an area has poor circulation. When there is reason to delay suturing in certain circumstances, such as a wound that is left open for drainage and then later closed, tertiary or delayed healing occurs.

Nursing process step: Assessment
Client needs category: Physiological integrity
Client needs subcategory: Physiological adaptation
Taxonomic level: Knowledge

38. *Correct answer:* **A**

Edges that are approximated are brought together, usually by sutures, tape, or staples. Wound edges that are erythematous and swollen are showing diffuse redness and enlargement. The death of areas of tissue or bone surrounded by healthy parts usually caused by insufficient blood supply is called necrosis. Wound edges that are gaping and draining are open with the free flowing of fluids, such as pus.

Nursing process step: Assessment
Client needs category: Physiological integrity
Client needs subcategory: Physiological adaptation
Taxonomic level: Knowledge

39. *Correct answer:* **C**

The reconstructive phase of wound healing takes longer and requires epithelization leading to healing by contraction and the formation of scar tissue. Collagen synthesis occurs during the proliferative phase of wound healing. Phagocytosis is the process involving microphages, which engulf microorganisms and cellular debris.

Nursing process step: Assessment
Client needs category: Physiological integrity
Client needs subcategory: Physiological adaptation
Taxonomic level: Knowledge

40. *Correct answer:* **D**

When an open wound is in the reconstructive phase, it must epithelize over a long period of time. If epithelization doesn't happen quickly enough, the wound becomes covered with eschar, which is dried protein and dead cells. Healing by primary intention occurs where the tissue surfaces have been approximated or closed. Necrotic and infected wounds may consist of purulent exudate, which consists of leukocytes, liquefied dead tissue debris, and dead and living bacteria. Suturing lacerations usually results in wound closure through primary intention.

Nursing process step: Implementation
Client needs category: Physiological integrity
Client needs subcategory: Physiological adaptation
Taxonomic level: Knowledge

41. *Correct answer:* **C**

Reformation of tissue at the site of injury requires protein, which can only be synthesized if sufficient amounts of amino acids are ingested. Carbohydrates and fat must be ingested to utilize protein for tissue repair rather than for basal metabolic activity. Vitamins are necessary for collagen synthesis and epithelization.

Nursing process step: Implementation
Client needs category: Physiological integrity
Client needs subcategory: Basic care and comfort
Taxonomic level: Knowledge

42. *Correct answer:* **B**

Debridement is the removal of all necrotic tissue and is necessary before beginning wound treatment. Debridement can be accomplished surgically by a doctor or through autolytic debridement, in which the wound is covered with an occlusive or semiocclusive dressing to provide a moist environment. Reparation of a wound can be accomplished by stapling wound edges or through sutures, which are threads used to sew body tissues together. Application of gauze dressings can be accomplished through several modes. Packing the wound with wet-to-wet gauze provides moisture, which helps dilute exudate but doesn't debride the wound. Packing the wound with wet-to-dry gauze promotes removal of necrotic debris. The debris is softened by the solution and adheres to the mesh gauze as it dries. The debris is then removed when the dressing is removed.

Nursing process step: Implementation
Client needs category: Physiological integrity
Client needs subcategory: Physiological adaptation
Taxonomic level: Cognitive: Knowledge

43. *Correct answer:* **C**

Purulent drainage consists of leukocytes, liquefied dead tissue debris, and dead and living bacteria. It indicates the presence of an infection and requires prompt treatment. Well-approximated edges indicate wound closure and are seen in primary or first intention healing. Granulation tissue occurs during the proliferative phase of wound healing, is pink to translucent red, and indicates the presence of blood supply, which brings oxygen and nutrients needed for tissue healing.

Nursing process step: Assessment
Client needs category: Physiological integrity
Client needs subcategory: Physiological adaptation
Taxonomic level: Knowledge

44. *Correct answer:* **B**

The placement of a padded footboard firmly against the patient's feet helps to prevent foot drop. Gatching the knee of the bed places the feet in a dependent position and won't assist in the prevention of foot drop. Placing a roll under the ankles may result in dorsiflexion and won't assist in the prevention of foot drop. Placing a trochanter roll under the thighs may assist with maintaining correct body alignment, but it won't assist in the prevention of foot drop.

Nursing process step: Implementation
Client needs category: Physiological integrity
Client needs subcategory: Reduction of risk potential
Taxonomic level: Application

45. *Correct answer:* **C**

Bed rest can result in muscle weakness, aches, atrophy, and disuse osteoporosis. Muscle strengthening, agility, and build-up result from increased muscle activity.

Nursing process step: Planning
Client needs category: Physiological integrity
Client needs subcategory: Reduction of risk potential
Taxonomic level: Comprehensive

46. *Correct answer:* **D**

Prolonged or intense pressure simultaneously produces ischemic changes and degeneration of the entire tissue segment between bony prominences and the skin surfaces. Pressure ulcers are most likely to occur over bony prominences, such as the sacrum, elbows, and heels.

Nursing process step: Assessment
Client needs category: Physiological integrity
Client needs subcategory: Reduction of risk potential
Taxonomic level: Comprehension

47. *Correct answer:* **D**

Passive range-of-motion exercises are performed to prevent shortening of muscles and tendons, deformities that limit joint function. Passive range-of-motion exercises may not strengthen muscles, but they don't prevent strengthening of muscles. These exercises may prevent contractures, but they can't restore loss of joint function. They may enhance circulation, but they don't overstimulate circulation.

Nursing process step: Planning
Client needs category: Physiological integrity
Client needs subcategory: Reduction of risk potential
Taxonomic level: Knowledge

48. *Correct answer:* **B**

When performing passive range-of-motion exercises, one should stop movement at the point of meeting resistance — not beyond, and never to the point of, pain. Passive range-of-motion exercises may cause fatigue, but they should not be performed to the point of fatigue.

Nursing process step: Implementation
Client needs category: Physiological integrity
Client needs subcategory: Reduction of risk potential
Taxonomic level: Application

49. *Correct answer:* **B**

The light-headedness and dizziness caused by rising after prolonged bed rest is due in part to muscular weakness and to orthostatic hypotension — or low blood pressure — related to the failure of arteriolar vasoconstriction. Venous stasis may result in dependent edema, but doesn't cause orthostatic hypotension. Orthostatic hypertension and orthostatic stasis are incorrect responses.

Nursing process step: Assessment
Client needs category: Physiological integrity
Client needs subcategory: Physiological adaptation
Taxonomic level: Comprehensive

50. *Correct answer:* **C**

Immobility can cause the patient to become dehydrated, which can lead to decreased blood viscosity, increased platelet counts, and increased blood coagulability. Stasis ulcers may occur with prolonged immobility, but they don't contribute to deep vein thrombosis (DVT). Increased anticoagulation medications can be used prophylactically to reduce the incidence of DVT in patients on prolonged bed rest and can also be used to therapeutically treat embolisms. The development of increased blood viscosity will enhance increased coagulability, possibly increasing the risk of DVT.

> *Nursing process step:* Planning
> *Client needs category:* Physiological integrity
> *Client needs subcategory:* Reduction of risk potential
> *Taxonomic level:* Comprehension

51. *Correct answer:* **A**

Constipation is a frequent complication of patients on bed rest due to many factors, such as dehydration, embarrassment in asking for the bedpan, the unnatural position for defecation, and decreased activity of the abdominal muscles, the diaphragm, and the levator ani. Diarrhea can be caused by certain medications, disease processes such as ulcerative colitis, and nutritional and malabsorptive disorders. Gastric reflux may occur as a result of pyloric stenosis or a motility disorder. Increased acid production may result from duodenal ulcers.

> *Nursing process step:* Planning
> *Client needs category:* Physiological integrity
> *Client needs subcategory:* Reduction of risk potential
> *Taxonomic level:* Comprehension

52. *Correct answer:* **A**

The patient should be positioned with the head of the bed completely flattened to perform an abdominal examination. If the head of the bed is elevated, the abdominal muscles and organs can be bunched up, altering the findings.

> *Nursing process step:* Assessment
> *Client needs category:* Health promotion and maintenance
> *Client needs subcategory:* Prevention and early detection of disease
> *Taxonomic level:* Knowledge

53. *Correct answer:* **B**

When assessing the abdomen, observation should precede auscultation. Auscultation is performed prior to palpation and percussion because palpation and percussion cause movement or stimulation of the bowel, which can increase bowel motility and thus intensify bowel sounds, creating false results.

Nursing process step: Assessment
Client needs category: Physiological integrity
Client needs subcategory: Physiological adaptation
Taxonomic level: Knowledge

54. *Correct answer:* **B**

Flatness is a soft, high-pitched, short sound produced by very dense matter, such as muscle. Tympany is a musical or drumlike sound percussed over an air-filled stomach. Resonance is a hollow sound produced over the lungs, and dullness is a thudlike sound percussed over dense tissue, such as the liver, spleen, or heart.

Nursing process step: Assessment
Client needs category: Health promotion and maintenance
Client needs subcategory: Prevention and early detection of disease
Taxonomic level: Knowledge

55. *Correct answer:* **A**

Guarding occurs when the patient is afraid palpation will cause pain and purposefully tightens his abdominal muscles. Experiencing pain in the right lower quadrant when the left lower quadrant is palpated is characteristically known as Rovsing's sign, a clinical manifestation of appendicitis. Intensification of pain when palpation is stopped is called rebound tenderness, which is also a clinical manifestation of appendicitis.

Nursing process step: Assessment
Client needs category: Physiological integrity
Client needs subcategory: Basic care and comfort
Taxonomic level: Comprehension

56. *Correct answer:* **D**

Jaundice occurs when the bilirubin concentration in the blood becomes abnormally increased. All body tissues, including the sclerae and the skin, become yellow-tinged or greenish-yellow. Liver failure results in the inability to excrete bilirubin, which results in jaundice. Jaundice isn't usually associated with acute renal failure, diabetes mellitus, or heart failure.

Nursing process step: Assessment
Client needs category: Physiological integrity
Client needs subcategory: Physiological adaptation
Taxonomic level: Comprehension

57. *Correct answer:* **B**

Icteric sclera is jaundice or a yellow tinge to the sclera. It can be of the extrahepatic type, which is caused by occlusion of the bile duct by a gallstone, or it can result from obstruction of the small bile ducts within the liver, known as intrahepatic obstruction. Both types can result in a buildup of bilirubin, which causes jaundice. Abdominal distention is usually associated with GI and rectal disorders. Low urine output may be associated with renal and cardiovascular disorders, and peripheral edema can be associated with cardiovascular and vascular disorders.

Nursing process step: Assessment
Client needs category: Physiological integrity
Client needs subcategory: Physiological adaptation
Taxonomic level: Comprehension

58. *Correct answer:* **D**

Body weight measurements taken at the same time each day in the same amount of clothing can provide a relatively accurate assessment of a patient's fluid status. Body temperature can be elevated in fluid volume deficit, but it may not demonstrate much variation in fluid volume excess. Anthropometric measurements include height, weight, skin-fold measurements, and mid-arm to upper-arm circumference. Skin-fold measurements are more indicative of the amount of body fat, the main form of stored energy. Blood pressure is the measurement of the pressure exerted by the blood as it flows through the arteries and can be influenced by many factors, such as age, exercise, stress, race, obesity, medications, and diurnal variations.

Nursing process step: Assessment
Client needs category: Physiological integrity
Client needs subcategory: Physiological adaptation
Taxonomic level: Comprehension

59. *Correct answer:* **A**

The side-lying position with the head of the bed lowered will help water and debris drain from the patient's mouth and prevent aspiration. Semi-Fowler's position describes a sitting position with the knees flexed. In a prone position, the patient lies on his abdomen with his head turned to one side; in the supine position, the patient lies flat on his back. The head and shoulders aren't elevated in the supine position.

Nursing process step: Implementation
Client needs category: Physiological integrity
Client needs subcategory: Reduction of risk potential
Taxonomic level: Comprehension

60. *Correct answer:* **A**

The duration of inner canthus to outer canthus is the natural tract for removal of debris from the eye; thus, the eye should be cleaned from inner canthus to outer canthus. This prevents debris and contaminants from being washed into the nasolacrimal duct. Washing the eye from outer to inner canthus, upper to lower canthus, or lower to upper canthus may result in debris being washed into the nasolacrimal duct, possibly interfering with proper drainage of tears or producing inflammation and swelling.

> *Nursing process step:* Implementation
> *Client needs category:* Physiological integrity
> *Client needs subcategory:* Reduction of risk potential
> *Taxonomic level:* Comprehension

61. *Correct answer:* B

If the blink reflex is lost or decreased and the eye is open, the air can dry the eye and cause a corneal abrasion or the development of exposure keratitis. Pupillary changes can result from increased intracranial pressure or medications. Conjunctival hemorrhage is due to the rupture of blood vessels, which can be caused by anything that causes bleeding within the body. Upper thoracic straining, such as forceful coughing or sneezing, may also result in conjunctival hemorrhage. Retinal detachment occurs when the neurosensory retina separates from the underlying pigment epithelium layer of the retina. Retinal detachment may be caused by illness, intraocular inflammation, trauma, or degenerative changes.

> *Nursing process step:* Implementation
> *Client needs category:* Physiological integrity
> *Client needs subcategory:* Reduction of risk potential
> *Taxonomic level:* Comprehension

62. *Correct answer:* A

Asepsis is the opposite of sepsis and means free from all pathogenic organisms. Medical asepsis includes all practices intended to confine a specific microorganism to a specific area, thereby limiting the number, growth, and transmission of organisms. Disinfection refers to the destruction or removal of pathogenic organisms.

> *Nursing process step:* Implementation
> *Client needs category:* Safe, effective care environment
> *Client needs subcategory:* Safety and infection control
> *Taxonomic level:* Knowledge

63. *Correct answer:* A

Sterilization is the process by which an object becomes free from *all* (not most) pathogenic organisms. This state can be obtained through the use of steam. Steam can be employed in two ways: as steam under pressure or as free steam. Autoclaves supply steam under pressure. Free steam is used to sterilize objects that would be destroyed at the higher temperature and pressure of an autoclave. Sterilization cannot be achieved by washing with soap or boiling in water.

Nursing process step: Implementation
Client needs category: Safe, effective care environment
Client needs subcategory: Safety and infection control
Taxonomic level: Comprehension

64. *Correct answer:* C

Sterile objects that are out of the line of vision are considered questionable, so it's necessary to keep the sterile field in one's line of vision at all times. The nurse's back should never be turned to the sterile field, nor should she reach across the sterile field and risk contaminating the field. Moving the sterile field away to reach for an object would risk contaminating the field.

Nursing process step: Implementation
Client needs category: Safe, effective care environment
Client needs subcategory: Safety and infection control
Taxonomic level: Application

65. *Correct answer:* D

Cold sterilization or soaking is used for objects with plastic or rubber parts that can't tolerate heat or for metallic objects prone to corrosion.

Nursing process step: Implementation
Client needs category: Safe, effective care environment
Client needs subcategory: Safety and infection control
Taxonomic level: Knowledge

66. *Correct answer:* D

Dry-heat sterilization is used to sterilize objects with sharp edges, needles, and some glass equipment to prevent the equipment from rusting or becoming dull.

Nursing process step: Implementation
Client needs category: Safe, effective care environment
Client needs subcategory: Safety and infection control
Taxonomic level: Knowledge

67. *Correct answer:* B

Moisture that passes through a sterile object draws microorganisms from non-sterile surfaces above and below the sterile surface. An impervious cover is a specially treated cover that is waterproof. Unless a sterile item has an impervious cover, the item becomes contaminated as soon as it's removed from the sterilizer. Equipment that has been tampered with, has a torn cover, or is open is considered contaminated.

Nursing process step: Implementation
Client needs category: Safe, effective care environment
Client needs subcategory: Safety and infection control
Taxonomic level: Knowledge

68. *Correct answer:* **C**

A sterile field is a work surface area prepared with sterile drapes to provide an area in which sterility is continually maintained. Clean towels and cloths are considered contaminated. Sterile packages leave areas of the work surface exposed and risk contaminating the field.

Nursing process step: Implementation
Client needs category: Safe, effective care environment
Client needs subcategory: Safety and infection control
Taxonomic level: Knowledge

69. *Correct answer:* **A**

When opening an envelope-wrapped sterile package, reaching across the package and using the first motion to open the top cover away from the body eliminates the need to later reach across the sterile field while opening the package. To remove equipment from the package, opening the first portion of the package toward, to the left, or to the right of the body would require reaching across a sterile field.

Nursing process step: Implementation
Client needs category: Safe, effective care environment
Client needs subcategory: Safety and infection control
Taxonomic level: Knowledge

70. *Correct answer:* **B**

The best way to place a sterile item on a sterile field when having to open a sterile package is to drop the item on the sterile field while ensuring that none of the packaging touches the sterile field. Placing the item on the field risks contamination by either the packaging or the nurse's hand. Handing the sterile package to a sterile nurse risks contaminating the sterile nurse with the packaging. Clean, gloved hands are considered contaminated.

Nursing process step: Implementation
Client needs category: Safe, effective care environment
Client needs subcategory: Safety and infection control
Taxonomic level: Knowledge

71. *Correct answer:* **C**

To prepare a sterile field while remaining sterile, one should open all the sterile containers (using aseptic technique) and place them on the work area. Next, put on sterile gloves and open the sterile drapes and towels, prepare the sterile field, and place the sterile items on the sterile field. Nonsterile containers should be opened prior to putting on sterile gloves. Opening unsterile containers with one gloved hand and preparing a sterile field with the other gloved hand risks contamination.

> *Nursing process step:* Implementation
> *Client needs category:* Safe, effective care environment
> *Client needs subcategory:* Safety and infection control
> *Taxonomic level:* Knowledge

72. *Correct answer:* **C**

The inner surface of the glove, including the part of the cuff turned over, will be in contact with the nurse's skin. The exterior of the gloves must remain sterile. Sliding a hand under the cuff touches a sterile area with an unsterile object and contaminates the area. Shaking the gloves from the wrap risks contamination. Picking up both gloves by the cuffs causes a sterile area to contact an unsterile area, resulting in contamination.

> *Nursing process step:* Implementation
> *Client needs category:* Safe, effective care environment
> *Client needs subcategory:* Safety and infection control
> *Taxonomic level:* Application

73. *Correct answer:* **A**

Having the fingers of the glove point down to the floor enables gravity to help the glove fingers stay open and helps the nurse to put on the glove. Having the fingers point in any other direction may result in the fingers folding over onto the glove, risking contamination.

> *Nursing process step:* Implementation
> *Client needs category:* Safe, effective care environment
> *Client needs subcategory:* Safety and infection control
> *Taxonomic level:* Application

74. *Correct answer:* **D**

Keeping sterile gloves sterile is of utmost importance. Anything below the waist is considered nonsterile. It's also important to keep sterile items in one's line of vision to ensure sterility. The hands may be kept folded, but they should be in one's line of vision.

Nursing process step: Implementation
Client needs category: Safe, effective care environment
Client needs subcategory: Safety and infection control
Taxonomic level: Application

75. *Correct answer:* **D**

The bottle rim and the inside of the cap are considered sterile. To ensure continued sterility of these parts of the bottle, one should place the cap face up and pour the liquid 6″ above the container into which the liquid is being poured. Placing the cap face down risks contaminating the cap, thus contaminating the entire solution once the cap is returned to the bottle.

Nursing process step: Implementation
Client needs category: Safe, effective care environment
Client needs subcategory: Safety and infection control
Taxonomic level: Knowledge

76. *Correct answer:* **D**

Once it touches the contaminated labia, the nondominant hand remains contaminated and should not touch anything sterile. Cleaning doesn't restore sterility to the nondominant hand. The urinary catheter must remain sterile to minimize the risk of urinary tract infection.

Nursing process step: Implementation
Client needs category: Safe, effective care environment
Client needs subcategory: Safety and infection control
Taxonomic level: Application

77. *Correct answer:* **C**

The urine in the patient's bladder and in the catheter tubing is considered sterile. The tubing should be clamped for a short time and then a needle with a syringe should be used to aspirate urine from the tubing. The urine should be placed in a sterile container. The urine in the collection bag is considered contaminated because the collection bag is frequently opened for emptying and could have been contaminated during the process.

Nursing process step: Implementation
Client needs category: Safe, effective care environment
Client needs subcategory: Safety and infection control
Taxonomic level: Application

78. *Correct answer:* **B**

Fastening the restraint to the bed frame on the same side of the bed prevents injuries when the side rails are placed up or down. Fastening the restraint to the bed frame on the opposite site of the bed risks injury with patient movement in the bed.

> *Nursing process step:* Implementation
> *Client needs category:* Safe, effective care environment
> *Client needs subcategory:* Safety and infection control
> *Taxonomic level:* Application

79. *Correct answer:* **B**

Restraints should be released from the patient's limbs every 2 hours to assess for skin breakdown, circulation, temperature, and limb motion. The patient should also be offered a drink and allowed bathroom privileges.

> *Nursing process step:* Implementation
> *Client needs category:* Safe, effective care environment
> *Client needs subcategory:* Safety and infection control
> *Taxonomic level:* Application

80. *Correct answer:* **A**

Chest restraints are used to prevent a patient from getting out of bed or falling out of a bed or a chair. A chest restraint should still allow for mobility within the bed, such as rolling on to one side or the other.

> *Nursing process step:* Implementation
> *Client needs category:* Safe, effective care environment
> *Client needs subcategory:* Safety and infection control
> *Taxonomic level:* Application

CHAPTER 6

PSYCHOSOCIAL NEEDS OF THE PATIENT

QUESTIONS

1. The mother of a 2-year-old patient tells the nurse that her child "cries and has a fit when I have to leave him with a sitter or someone else." Which of the following statements would be the nurse's most accurate analysis of the mother's comment?

    A. The child has not experienced limit-setting or structure.
    B. The child is expressing a physical need, such as hunger.
    C. The mother has nurtured overdependence in the child.
    D. The mother is describing her child's separation anxiety.

2. In a weekly parenting class, the nurse teaches parents ways to foster healthy self-concepts in their children. Which method is most important?

    A. Offering clear guidelines
    B. Setting fluctuating limits
    C. Providing frequent correction
    D. Giving positive feedback

3. During a parenting group, a mother reveals that she has observed her six-year-old son playing with his penis during his bath. She asks the nurse how she should respond to the situation. What would be the nurse's most appropriate response?

    A. "Avoid characterizing it as bad because that's normal at his age."
    B. "If he does it often, you might consider disciplining him."
    C. "Ignore it and hope that he outgrows it."
    D. "Supervise his baths and encourage him to stop doing it."

4. Margaret Miller has recently begun menopause. On her initial visit to the clinic, she expresses dejection about having no children despite having been pregnant three times. In planning Mrs. Miller's care, it would be most therapeutic if the nurse focuses on which need?

    A. Esteem
    B. Love and belonging
    C. Physiological
    D. Safety

5. Autumn Springer, a 15-year-old girl with scoliosis, has recently received an invitation to a swim party. She asks the nurse how she can disguise her impairment when dressed in a bathing suit. Which nursing diagnosis can be best justified by Autumn's statement?

   A. Anxiety
   B. Body image disturbance
   C. Ineffective individual coping
   D. Social isolation

SITUATION: *Jeremy Derstein, age 6, is the son of a hemophiliac who has end-stage acquired immune deficiency syndrome (AIDS). The father is being seen by the nurse for treatment of severe oral candidiasis.*

   Questions 6 to 8 relate to this situation.

6. During a clinic visit, Jeremy asks the nurse what it will be like when his father dies. In discussing death with Jeremy, which of the following would be a most therapeutic expression for the nurse to use?

   A. "He'll be going to sleep, Jeremy."
   B. "He will die, Jeremy, and you and I will miss him."
   C. "It's still possible for your dad to get better, Jeremy."
   D. "Jeremy, your dad will be going away."

7. In caring for a child who exhibits the manifestations of spiritual distress, the nurse knows that the most likely cause of spiritual distress is:

   A. arrival of a new sibling.
   B. change in living situation.
   C. loss of a caretaker.
   D. taking on new responsibilities.

8. In discussing death with Jeremy's family, it would be most helpful if the nurse understands that Jeremy's 4-year-old brother most likely perceives death in which way?

   A. An insignificant event unless taught otherwise
   B. Punishment for something the individual did
   C. Something that just happens to older people
   D. Temporary separation from the loved one

SITUATION: *Cecily Ray is a 15-year-old patient on a hematology unit who is being treated for sickle cell crisis.*

   Questions 9 to 16 relate to this situation.

9. During a crisis such as that seen in sickle cell anemia, aldosterone release is stimulated. In what way might this influence Cecily's fluid and electrolyte balance?

A. Sodium loss, water loss, and potassium retention
B. Sodium loss, water retention, and potassium loss
C. Sodium retention, water loss, and potassium retention
D. Sodium retention, water retention, and potassium loss

10. Stress increases the secretion of a number of endogenous hormones. Which increased hormonal secretion most likely contributes to the worsening of Cecily's sickle cell crisis?

A. Antidiuretic hormone
B. Estrogen
C. Glucagon
D. Norepinephrine

11. During the crisis, Cecily receives frequent pain medication. Which method of administration will most help reduce her sense of powerlessness?

A. Intramuscular injection
B. I.V. injection
C. Oral ingestion
D. Patient-controlled device

12. In assessing Cecily's psychosocial development, which observation by the nurse would most suggest that Cecily has a developmental lag?

A. Complaints that her parents are too strict
B. Frequent requests for back massages
C. No visits from friends her age
D. A request to keep her curtain closed at all times

13. In planning Cecily's nursing care, which of the following statements or questions by the nurse would most effectively assess Cecily's self-esteem?

A. "Are you happy with yourself and your life in general?"
B. "Share some of your likes and dislikes with me."
C. "Tell me about your sense of satisfaction with yourself."
D. "What changes in your body most concern you?"

14. According to Maslow's hierarchy of needs, which priority are the individual's esteem needs?

A. Second
B. Third
C. Fourth
D. Fifth

15. During her hospitalization, Cecily has visitors who read religious literature to her. Her roommate complains that she does not agree with what the visitors are teaching, and she wants it to stop. In addressing this concern, which concept is most important for the nurse to understand?

    A. A nurse should ensure that the spiritual needs of one patient don't supersede the privacy needs of another.

    B. A person has a right to pursue her convictions when they don't infringe on the rights of another.

    C. In a health care setting, the emphasis should focus on the physical needs of the patients involved.

    D. Individual freedoms can be tolerated only when they don't offend other individuals.

16. The doctor discusses the possibility of a blood transfusion with Cecily and her family. Shortly afterward, Cecily exhibits mild anxiety and states, "If I get a blood transfusion, I'll get AIDS." In managing Cecily's anxiety in this situation, which concept should most guide the nurse?

    A. Anxiety distorts reality.

    B. Dispelling a myth dispels anxiety.

    C. Information decreases anxiety.

    D. Providing privacy has priority.

SITUATION: *Benjamin Tizer is a 17-year-old high school student on an orthopedic unit following a motorcycle accident. He is in skeletal traction to repair a compound fracture of his right femur.*

Questions 17 to 22 relate to this situation.

17. Benjamin's nurses have noticed a recent change in his behavior, including using poor eye contact during interactions, refusing to participate in his own care, and complaints that "no one really wants me to get better." The nurse identifies his nursing diagnosis as hopelessness. Which is the most appropriate related goal? The patient will:

    A. ask to speak to a member of the clergy or a social worker.

    B. demonstrate cheerfulness and contentment.

    C. request literature on self-help and motivation.

    D. shave himself and brush his teeth at least once a day.

18. Benjamin is an orthodox Jew. Which food is most important to exclude from his diet?

A. Aged cheeses
B. Lamb or mutton
C. Shellfish
D. Yeast breads

19. After learning that his skeletal traction needs to continue for two weeks beyond the original plan, Benjamin becomes severely anxious. Which physical manifestations most support the nurse's diagnosis of severe anxiety?

A. Elevated respiratory and heart rates
B. Choking and chest pain
C. Pupil dilation and dry mouth
D. Tremors, sweating, and nausea

20. Which nursing intervention would be most relevant to treat Benjamin's severe anxiety?

A. Ask him to select a calming activity, such as reading or listening to music.
B. Ask him what caused the anxiety and what has helped in the past.
C. Gently instruct him to calm down and try to be still.
D. Take measures to make his environment less distracting.

21. Benjamin's "as-needed" medication list includes several drugs. Which of the following drugs is one of the most widely used antianxiety agents?

A. Diphenhydramine (Benadryl)
B. Flumazenil (Romazicon)
C. Lorazepam (Ativan)
D. Methylphenidate (Ritalin)

22. The nurse teaches Benjamin progressive muscle relaxation as a means of stress reduction. Which statement by Benjamin most suggests he has been properly performing the technique?

A. "During the technique, I look at the muscles that I am working with at the time."
B. "I hold my breath while I am tensing and relaxing each muscle group and then I breath normally."
C. "I keep each muscle group tightened until the point I begin sensing pain and then I relax it."
D. "When I experience muscle cramping, I stop the technique and massage the muscle that's cramped."

SITUATION: *Gill Martin, a 21-year-old college student, fell from a train trestle and sustained a spinal cord injury, leaving him paralyzed below the waist. He's in a spinal cord rehabilitation program and is refusing to do things for himself or participate in his prescribed program.*

Questions 23 to 27 relate to this situation.

23. Which developmental stage most represents the psychosocial challenge associated with Mr. Martin's age-group?

    A. Identity versus role confusion
    B. Industry versus inferiority
    C. Integrity versus despair
    D. Intimacy versus isolation

24. One of Mr. Martin's nursing diagnoses is self-esteem disturbance. Which of the following is the most therapeutic nursing intervention?

    A. Asking his friends to encourage his self-care
    B. Enlisting him in the planning of his own care
    C. Moving him to a different hospital environment
    D. Teaching him how to perform self-care measures

25. The nurse says to Mr. Martin, "Tell me about your plans after hospitalization." Mr. Martin replies, "I think in a few months I'll pick up where I left off—back in college doing what I was doing before." Which stage of the grief-loss process is Mr. Martin demonstrating?

    A. Acceptance
    B. Anger
    C. Bargaining
    D. Denial

26. During his admission, Mr. Martin has frequent visits from a male friend. At one point the nurse enters the room and finds him and his friend holding hands. Mr. Martin states, "I know what you're thinking and you're wrong." Which concept should guide the nurse in her response to Mr. Martin? The nurse needs to:

    A. help the patient work through an identity crisis.
    B. foster a healthy self-esteem in her patient.
    C. disguise any rejection of the behavior she saw.
    D. project an objective and nonjudgmental attitude.

27. The nurse finds Mr. Martin attempting to move his legs and crying. Mr. Martin says, "Please pray with me that I can walk again." Which would be the most therapeutic response from the nurse?

    A. "Let's just have a moment of silence."
    B. "How would you like us to pray?"
    C. "Perhaps I can call the chaplain."
    D. "You pray and I'll stay with you."

SITUATION: *Sara Livingston, a 31-year-old mother of three, is a patient on a burn unit. She received full-thickness burns on 20% of her body in a house fire in which two of her children died.*

Questions 28 to 34 relate to this situation.

28. During her hospitalization, Ms. Livingston is placed on a routine I.V. histamine antagonist because of the risk of a stress ulcer (Curling's ulcer). Which stress-related physiologic changes in the upper GI tract predispose Ms. Livingston to a stress ulcer?

    A. Blockage of the stomach's parietal cells
    B. Decreased vagal nerve stimulation
    C. Elevation of the gastric juice pH
    D. Lessening of gastric mucosa production

29. Which of the following drugs listed on Ms. Livingston's medication administration record is a histamine antagonist?

    A. Ciprofloxacin (Cipro)
    B. Morphine sulfate (Morphine)
    C. Ranitidine hydrochloride (Zantac)
    D. Sucralfate (Carafate)

30. Which behavior would most suggest to the nurse that Ms. Livingston is still in the earliest stage of the grief process?

    A. Outbursts of anger toward her family and the staff
    B. Questions about job retraining
    C. Statements that "it's a dream" and "it didn't really happen"
    D. Wanting to be left alone in a dark and quiet room

31. Shortly after midnight, Ms. Livingston is awakened by the sound of an arriving ambulance outside the window of her room. She exhibits screaming, crying, vigorous attempts to get out of bed, and incoherent speech. These manifestations are most suggestive of which level of anxiety?

    A. Mild
    B. Moderate
    C. Panic
    D. Severe

32. During this episode, which nursing action is most important?

    A. Discuss appropriate coping mechanisms with Ms. Livingston.

    B. Encourage Ms. Livingston to express her feelings about the event.

    C. Have Ms. Livingston remain in bed and apply soft wrist restraints.

    D. Stay with Ms. Livingston and provide assurance and safety.

33. During a hydrotherapy session for treatment of Ms. Livingston's burns, the nurse notices the patient is calmer and more lucid in her interaction. Which of the following statements would be the most supportive rationale for the patient's behavior?

    A. Submersion in the water of the hydrotherapy enables Ms. Livingston to ignore her wounds.

    B. The warmth and whirlpool action of hydrotherapy distracts Ms. Livingston from her physical and psychic discomfort.

    C. The hydrotherapy is numbing Ms. Livingston's peripheral sensory nerves, allowing her to relax.

    D. While away from the environment of the hospital room, Ms. Livingston has more liberty to relax and converse.

34. During the therapy, Ms. Livingston exhibits spiritual distress. Which statement by the patient most supports this nursing diagnosis?

    A. "I haven't been to church since I was a child."

    B. "I must be being punished for something I did."

    C. "I'm very sad and I miss my little ones terribly."

    D. "I want my children to have a nonreligious burial."

SITUATION: *The Pombo family has requested a home visit from a community health nurse because two of their four children have recently been injured in the home.*

Questions 35 to 37 relate to this situation.

35. As the nurse is discussing details of the children's injuries, she observes Mr. Pombo's use of several defense mechanisms. Which statement by him demonstrates his use of projection?

    A. "Every child has bumps and bruises; everybody knows that."

    B. "Look, nurse—what goes on in this house is none of your business."

    C. "The injuries are not my doing, they're my wife's."

    D. "The kids aren't injured—those are normal childhood bruises."

36. When the nurse addresses his use of displacement, Mr. Pombo asks her to define the term. The nurse would be most accurate if she defined "displacement" as:

    A. aiming one's true emotions or response at a more vulnerable or safer target than the target for which they are intended.

    B. avoiding the acceptance of responsibility for an event or behavior that is obviously the individual's responsibility.

    C. behaving at a level of maturity that is lower than the level which the individual has previously attained.

    D. fabricating seemingly logical or plausible reasons for a particular unacceptable behavior.

37. In teaching the Pombos about fostering a healthy self-concept in their children, the key concept the nurse teaches is to:

    A. give their children free reign in self-expression.

    B. provide clearly defined standards and limits.

    C. supervise their children's interaction with friends.

    D. withhold praise until potential has been met.

SITUATION: *Marianne Elayyan, a parish nurse, conducts socialization groups for the elderly, health screening clinics, and home visits.*

Questions 38 to 42 relate to this situation.

38. If Mrs. Elayyan prioritizes her programs according to Maslow's hierarchy of basic human needs, on which effort would she work first?

    A. Organizing a group of volunteers to run a monthly food bank and assist at the blood bank

    B. Presenting a speakers' bureau on topics such as role adjustment and stress management

    C. Sponsoring potlucks, dances, and singles mixers for the seniors in the community

    D. Teaching a class on scheduling and organizing accurate medication usage to prevent underdosing or overdosing

39. In conducting her programs, which of the following concepts should Mrs. Elayyan realize is a spiritual need common to all people?

    A. Faithfulness

    B. Forgiveness

    C. Peacefulness

    D. Trustworthiness

40. In teaching during her health-screening clinic, which information about the physiologic effects of stress would Mrs. Elayyan be most accurate in providing? Stress:

    A. brings about an overall improvement in digestion.
    B. can inhibit one's ability to fight off infection.
    C. decreases the body's metabolism and calorie usage.
    D. has no benefit and should be avoided altogether.

41. Which patient's behavior most suggests to Mrs. Elayyan that self-actualization has been obtained? The patient who:

    A. consistently involves herself in projects aimed at helping others.
    B. continually looks for new experiences and adventure in life.
    C. declares, "I'm a good person who has never harmed anyone."
    D. has successfully raised a family and is retired comfortably.

42. During a home visit, the patient tells Mrs. Elayyan that his spinal arthritis keeps him from engaging in sexual intercourse with his wife. He tells her, "We're nowhere as close as we used to be because of it." Which is the most therapeutic response from Mrs. Elayyan?

    A. "I understand your concerns, and it's something we can eventually talk about."
    B. "Might you be using the arthritis as an excuse for not being intimate with your wife?"
    C. "Tell me more about how things have changed between you and your wife."
    D. "Try different ways of gratification, perhaps different positions or techniques."

SITUATION: *Consuela Rivas is an occupational health nurse for a large petroleum plant. She conducts a weekly stress management clinic.*

    Questions 43 to 50 relate to this situation.

43. While Miss Rivas is teaching a stress management class, a patient who has been reading literature on stress states, "The book I'm reading keeps mentioning 'endorphins.' What are endorphins?" The most accurate response from Miss Rivas would be that endorphins are:

    A. a family of stress-blocking hormones in the body.
    B. catalysts that prevent the symptoms of stress.
    C. enzymes that mediate the body's stress response.
    D. pain-blocking chemicals produced by the body.

44. Which of the following choices depletes the body's endorphin levels?

    A. Brief pain

    B. Physical exercise

    C. Recurrent stress

    D. Sexual activity

45. Which set of vital signs in a normal healthy adult most suggests that an individual is moderately anxious?

    A. Temperature 98.6°F (37°C); pulse 80; respirations 16; blood pressure 120/80

    B. Temperature 99.2°F (37.3°C); pulse 104; respirations 22; blood pressure 136/90

    C. Temperature 97.4°F (36.3°C); pulse 66; respirations 12; blood pressure 106/76

    D. Temperature 101.4°F (38.6°C); pulse 120; respirations 8; blood pressure 90/60

46. A patient has non–insulin-dependent diabetes mellitus and notices an unusual variance in his fingerstick blood sugar during periods of stress. Which of the following most accurately explains this variance?

    A. Stress causes peripheral vasoconstriction, which alters the accuracy of fingerstick blood sugar values.

    B. Stress consumes energy reserves and increases insulin production, therefore reducing blood sugar.

    C. Stress decreases insulin release and mobilizes glucose reserves, which elevates fingerstick blood sugars.

    D. Stress provokes wide fluctuations in blood sugar levels because of elevated glucagon and insulin levels.

47. When describing anxiety to her stress management class, which statement by Miss Rivas would be most accurate when she is discussing the benefits of mild anxiety?

    A. It increases one's ability to go to sleep and stay asleep.

    B. It facilitates the body's storage of energy.

    C. It improves visual and auditory acuity.

    D. It focuses an individual's attention to details.

48. A patient of Miss Rivas's recently became blind from an industrial accident and has been receiving therapy to help him work through the loss and grief process. Which behavior by the patient would most suggest that the therapy is having its intended effect?

    A. Turning to one's family, friends, and caregivers for support
    B. Expressing certainty that he'll eventually recover his sight
    C. Purchasing a number of books on audiotape and a pair of sunglasses
    D. Setting a date to return to the job he held when injured

49. A patient reveals to Miss Rivas that he's sexually impotent. Which statement best expresses what the nurse should understand her patient to mean?

    A. He's disinterested in sexual intimacy.
    B. He's unable to attain or retain an erection.
    C. His ejaculate is void of sperm.
    D. His semen isn't potent enough to impregnate.

50. In planning a support group for individuals identified as at risk for myocardial infarction due to work-related stress, which employee group should Miss Rivas focus on the most?

    A. Those with a high level of independence in decision making
    B. Those with minimum workplace autonomy and control
    C. Those with project deadlines and busy appointment schedules
    D. Those with schedules involving evening and weekend shifts

SITUATION: *Keyshaun Prichett, a 5-year-old kindergarten student, is being seen in a pediatric clinic for routine childhood immunizations.*

Questions 51 to 53 relate to this situation.

51. Which statement about Keyshaun most demonstrates a positive self-concept?

    A. She easily submits to the nurse giving her the immunization.
    B. She pays careful attention when told what will be done.
    C. She receives high praise from her mother during the visit.
    D. She states that she knows the alphabet and can count to 50.

52. While a nurse is preparing the immunizations, Keyshaun's mother says, "Keyshaun is just getting these drugs because the school requires it. God will protect her from illness." Which is the most appropriate response from the nurse?

    A. "Before I give the immunizations, would you like to talk about it?"
    B. "In a way you're right — God uses medicine to prevent illness."
    C. "Really it's not the school that requires it; it's the government."
    D. The nurse listens to the comment, but does not verbally respond.

53. Avoidance of which behavior by Keyshaun would most suggest that she isn't succeeding in achieving the developmental task of her age-group?

    A. Identity
    B. Initiative
    C. Integrity
    D. Intimacy

SITUATION: *Melissa Alba, a 20-year-old mother of a premature newborn, smoked cigarettes during her pregnancy. Her son is a patient in a neonatal intensive care unit and has a diagnosis of acute respiratory distress syndrome.*

Questions 54 to 57 relate to this situation.

54. Ms. Alba is expressing guilt about her son's illness. Which aspect of her role should the nurse most express when addressing Ms. Alba's guilt?

    A. Empathy
    B. Guidance
    C. Role modeling
    D. Teaching

55. Because Ms. Alba is Roman Catholic, which nursing intervention would be most appropriate for the nurse to discuss with her?

    A. Baptism of the infant
    B. Circumcision of the infant
    C. Last rites for Ms. Alba
    D. Sacraments of the sick for Ms. Alba

56. In an interaction with Ms. Alba's mother, a nurse learns that the infant's father recently committed suicide and indicated Ms. Alba as the cause of his suicide. When the nurse asks Ms. Alba who will help her when she takes her child home from the hospital, Ms. Alba displays her use of the defense mechanism of repression. Which of her comments depicts repression?

    A. "My boyfriend, the baby's father, will be there when I need him."
    B. "I don't expect to need any help from anyone when he comes home."
    C. "I can't take him home until he's completely well—no one can help."
    D. "Bring my baby home? I'll lose him just like I did my boyfriend."

57. Which behavior by the neonatal nurses would be most beneficial in helping Ms. Alba's son meet the developmental landmark of infancy?

    A. Allowing the infant long periods of sleep without interruption
    B. Keeping the infant in a warm and clean isolette
    C. Meeting the infant's needs with consistency and warmth
    D. Placing the infant on a demand feeding schedule

SITUATION: *Bruce Denton, a 23-year-old homeless man, is seen by the community health nurse in the sexually transmitted disease clinic. He has recently noticed skin discoloration on his palms, face, and chest.*

Questions 58 to 62 relate to this situation.

58. Although it's clearly documented that Mr. Denton has contracted human immunodeficiency virus (HIV) using unclean I.V. needles, he denies it in an interaction with the community health nurse. Which nursing intervention would be most therapeutic in focusing on Mr. Denton's denial?

    A. Give him specific feedback in the areas where denial exists.
    B. Ignore the patient's use of denial because he has a terminal illness.
    C. Provide literature that shows the relationship between I.V. drugs and HIV infection.
    D. Refer the patient to a clinical nurse specialist in infection control.

59. Mr. Denton develops full-blown acquired immunodeficiency syndrome and expresses to the nurse that he "will do anything to not die from this." In which phase of the dying process is Mr. Denton?

    A. Acceptance
    B. Anger
    C. Bargaining
    D. Depression

60. Mr. Denton asks the community health nurse to shave his face because he "doesn't want to look in a mirror." Which nursing diagnosis most encompasses this statement?

    A. Altered health maintenance
    B. Body image disturbance
    C. Ineffective individual coping
    D. Self-care deficit

61. What is the most appropriate nursing response to Mr. Denton's request for the nurse to shave him?

A. Gather the supplies and remind him shaving is something he can do himself.
B. Help him shave and use the opportunity to build rapport with him.
C. Suggest he postpone shaving until he is ready to look at himself.
D. Tell him that before others can accept him, he must accept himself.

62. Because Mr. Denton has a sexually transmitted disease, it's important for the community health nurse to address his sexuality. Which concept should most direct the nurse in dealing with this concern?

A. Sexuality can have both positive and negative effects on the community in general.
B. Sexuality is on a continuum ranging from normal sexuality to aberrant sexuality.
C. Sexuality is personal and should only be addressed when directed by the patient.
D. Sexuality should be addressed with patients only when there is a suspected problem.

SITUATION: *Stanley Gregorski, a 28-year-old law student and recent newlywed, is a patient in a surgical oncology unit following a bilateral orchiectomy for testicular cancer.*

Questions 63 to 69 relate to this situation.

63. Mr. Gregorski has postoperative anxiety and a pain level of 9 on a 10-point pain scale. Which drug would be most effective in treating both of these symptoms?

A. Diazepam (Valium)
B. Morphine sulfate (Morphine)
C. Phenobarbital (Luminal)
D. Prochlorperazine (Compazine)

64. Which of the following choices is the most common adverse effect of anxiolytic drugs?

A. Cardiac arrhythmias
B. Drowsiness and confusion
C. Respiratory depression
D. Subnormal blood pressure

65. Mr. Gregorski asks the nurse if he can still be a father. Which response from the nurse is most useful in building Mr. Gregorski's self-concept?

    A. "There are many ways a man can be a father even if he's sterile."
    B. "This surgery has not taken away your ability to be a great father."
    C. "Yes, you're a wonderful husband and will be a wonderful father."
    D. "Your testicles were removed, so you can't get your wife pregnant."

66. Which of the following choices gives the best evidence that Mr. Gregorski is coping with an altered body image?

    A. A nurse overhears him tell a friend, "I'm the same man I've always been."
    B. He makes plans to join a health club and begin an exercise program.
    C. He prefers not to talk about the surgery and wants to forget about it.
    D. His wife says their intimacy was not affected by his surgery.

67. In discussing with Mr. Gregorski the effect of certain drugs on sexual performance, which endogenous neurohormone should the nurse know is most responsible for mobilizing sexual behavior?

    A. Androgen
    B. Dopamine
    C. Serotonin
    D. Testosterone

68. Mr. Gregorski's spiritual guidance is derived from a Protestant perspective; therefore, it's most accurate for the nurse to realize that he might attribute his illness to which spiritual concept?

    A. Chance or fate
    B. Lack of faith
    C. Punishment
    D. The will of God

69. Mr. Gregorski asks the nurse, "How am I going to discuss my illness and its consequences with my wife?" The nurse invites Mr. Gregorski to role play with her. The nurse would most likely:

    A. act out how Mr. Gregorski should inform his wife.
    B. coach Mr. Gregorski in the most effective way to talk to his wife.
    C. present a series of scenarios and discuss them with Mr. Gregorski.
    D. simulate the spousal interaction with Mr. Gregorski.

SITUATION: *Shakira Fahim, a 33-year-old police officer, is being seen in the chronic pain clinic because she injured her back during an arrest. She has been having sciatica for 4 months, and comfort-relieving therapies have been unsuccessful.*

Questions 70 to 72 relate to this situation.

70. Which of the following statements by the nurse would be most useful in assessing the major psychosocial developmental tasks of Ms. Fahim's age-group?

  A. "Look back on your life for me and give me an overview."
  B. "Have you started planning your family and when it will start?"
  C. "Tell me about your work and the people important in your life."
  D. "What were some of the troubling experiences of your childhood?"

71. Failure to achieve the developmental landmark for Ms. Fahim's age-group would be evident by which behavior?

  A. Abandoning long-held beliefs
  B. Changing employers every two years
  C. Having difficulty maintaining friendships
  D. Worrying about never having children

72. Ms. Fahim is a black Muslim. Which of the following choices is a concept on which black Muslims focus?

  A. Celibacy
  B. Prophecy
  C. Reincarnation
  D. Self-esteem

SITUATION: *Soon Nguyen, a 46-year-old recent immigrant to the United States, is hospitalized on a respiratory isolation unit for multiple drug-resistant pulmonary tuberculosis.*

Questions 73 to 75 relate to this situation.

73. What is the most appropriate way in which the nurse can meet Mrs. Nguyen's love and belonging needs?

  A. Include family in her care.
  B. Intervene with frequent touch.
  C. Offer rest and comfort measures.
  D. Provide a phone in the room.

74. Because of the risk of microbe transmission from pulmonary secretions, sexual intimacy is discouraged while Mrs. Nguyen is isolated. To accommodate Ms. Nguyen's sexual needs, which action would be most appropriate from the nurses?

    A. Encouraging her patience
    B. Giving daily massages
    C. Listening to her concerns
    D. Providing periods of privacy

75. Mrs. Nguyen belongs to a charismatic denomination and believes in practicing "the gifts of the spirit." Which of the following behaviors would be most associated with her religious convictions?

    A. Asking to have only female caregivers
    B. Refusing caffeinated beverages and chocolate
    C. Wanting to receive a laying on of hands
    D. Wearing coverings and clothing considered sacred

SITUATION: *Carla Brown, a 53-year-old mortgage banker, is admitted to an emergency department with chest pain and dyspnea.*

Questions 76 to 79 relate to this situation.

76. Which of the following choices is the developmental milestone Ms. Brown's age-group is most likely seeking to achieve?

    A. Addressing unresolved conflicts
    B. Maintaining a youthful persona
    C. Making a contribution to the world
    D. Reconsidering career choices

77. According to Erikson's stages of development, which term would most identify the failure of Ms. Brown to master the developmental crisis of her stage?

    A. Despair
    B. Isolation
    C. Role confusion
    D. Stagnation

78. Which of the following would most likely decrease as a result of the pain and anxiety experienced from angina and admission to an emergency department?

A. Bodily oxygen demand
B. Bronchial lumen diameter
C. Diameter of the pupils
D. Urinary bladder contractions

79. Ms. Brown is a Mormon and wears a special garment that has a distinctive meaning to Latter-day Saints. Which information about the garment would most accurately assist the nurse in planning Ms. Brown's nursing care?

A. It's never to be seen by non–family members.
B. It's to be worn at all times.
C. The nurse should avoid touching the garment.
D. The patient should not be asked about it.

SITUATION: *Brian Watson, a 54-year-old disabled auto mechanic, has chronic hypertension and insulin-dependent diabetes mellitus. He has been receiving hemodialysis three times a week for three years and is on the kidney transplant waiting list.*

Questions 80 to 82 relate to this situation.

80. Which behavior by Mr. Watson gives the most evidence that he is experiencing hopelessness?

A. Absence of his typical vigor for life
B. A declaration that he's an atheist
C. His fear he won't receive a transplant
D. Rare visits from friends and family

81. When assessing Mr. Watson's role performance, it's most important for the nurse to gather information about his:

A. family size and the family dynamics.
B. individual perceptions about his role.
C. occupational interests and job history.
D. overview of how he spends his time.

82. Which response from the nurse is most therapeutic when Mr. Watson expresses a spiritual point of view with which the nurse disagrees?

A. "Can we talk about your plans when you leave the hospital?"
B. "That sounds out of the mainstream, Mr. Watson."
C. The nurse actively listens to Mr. Watson.
D. "Yes, Mr. Watson, I respect your spiritual point of view."

SITUATION: *Nicholas Eleftherakis, a 68-year-old retired tobacco farmer, is terminally ill with metastatic lung cancer and is being cared for by his family at home. Hospice nurses visit him three times a week.*

Questions 83 to 86 relate to this situation.

83. Mr. Eleftherakis's daughter says to a nurse, "None of us can get along without Dad. He can't die — we need him too badly." Which core concept of grief work should guide the nurse in her response to the daughter?

    A. Denial is a normal stage in the grieving process and should be promoted by the nurse.
    B. Grief work most appropriately begins after the death of the loved one, not during or before his death.
    C. Showing sympathy toward the patient and family enables the nurse to be most effective.
    D. The needs met by key people in our lives can be met in other ways and by other people.

84. In meeting Mr. Eleftherakis's needs, which of the following is most important for the nurse to communicate?

    A. Advocacy
    B. Caring
    C. Competence
    D. Organization

85. Early in Mr. Eleftherakis's illness, his family experiences role conflict. Role conflict is best defined as a situation in which:

    A. a person assumes a new role with which he is inexperienced.
    B. a role requires an individual to experience discord with others.
    C. expectations of one role of an individual compete with expectations of another of the individual's roles.
    D. one person assuming the role of another causes a disturbance.

86. During a conversation with the hospice nurse, Mr. Eleftherakis says, "I've worked all my life, taken pretty good care of my family, and now I'm going to die and go to who-knows-where." Which of the following choices is the most appropriate response from the nurse?

    A. "Cheer up, Mr. Eleftherakis, you're still here with your family."
    B. "Death is different for all, and no one knows what to expect."
    C. "I will be glad to contact a chaplain who can talk to you."
    D. "You're concerned about your death and what will happen."

ANSWER SHEET

	A B C D		A B C D		A B C D		A B C D
1	○ ○ ○ ○	23	○ ○ ○ ○	45	○ ○ ○ ○	67	○ ○ ○ ○
2	○ ○ ○ ○	24	○ ○ ○ ○	46	○ ○ ○ ○	68	○ ○ ○ ○
3	○ ○ ○ ○	25	○ ○ ○ ○	47	○ ○ ○ ○	69	○ ○ ○ ○
4	○ ○ ○ ○	26	○ ○ ○ ○	48	○ ○ ○ ○	70	○ ○ ○ ○
5	○ ○ ○ ○	27	○ ○ ○ ○	49	○ ○ ○ ○	71	○ ○ ○ ○
6	○ ○ ○ ○	28	○ ○ ○ ○	50	○ ○ ○ ○	72	○ ○ ○ ○
7	○ ○ ○ ○	29	○ ○ ○ ○	51	○ ○ ○ ○	73	○ ○ ○ ○
8	○ ○ ○ ○	30	○ ○ ○ ○	52	○ ○ ○ ○	74	○ ○ ○ ○
9	○ ○ ○ ○	31	○ ○ ○ ○	53	○ ○ ○ ○	75	○ ○ ○ ○
10	○ ○ ○ ○	32	○ ○ ○ ○	54	○ ○ ○ ○	76	○ ○ ○ ○
11	○ ○ ○ ○	33	○ ○ ○ ○	55	○ ○ ○ ○	77	○ ○ ○ ○
12	○ ○ ○ ○	34	○ ○ ○ ○	56	○ ○ ○ ○	78	○ ○ ○ ○
13	○ ○ ○ ○	35	○ ○ ○ ○	57	○ ○ ○ ○	79	○ ○ ○ ○
14	○ ○ ○ ○	36	○ ○ ○ ○	58	○ ○ ○ ○	80	○ ○ ○ ○
15	○ ○ ○ ○	37	○ ○ ○ ○	59	○ ○ ○ ○	81	○ ○ ○ ○
16	○ ○ ○ ○	38	○ ○ ○ ○	60	○ ○ ○ ○	82	○ ○ ○ ○
17	○ ○ ○ ○	39	○ ○ ○ ○	61	○ ○ ○ ○	83	○ ○ ○ ○
18	○ ○ ○ ○	40	○ ○ ○ ○	62	○ ○ ○ ○	84	○ ○ ○ ○
19	○ ○ ○ ○	41	○ ○ ○ ○	63	○ ○ ○ ○	85	○ ○ ○ ○
20	○ ○ ○ ○	42	○ ○ ○ ○	64	○ ○ ○ ○	86	○ ○ ○ ○
21	○ ○ ○ ○	43	○ ○ ○ ○	65	○ ○ ○ ○		
22	○ ○ ○ ○	44	○ ○ ○ ○	66	○ ○ ○ ○		

ANSWERS AND RATIONALES

1. *Correct answer:* **D**

Before coming to any conclusion, the nurse should ask the mother focused questions; however, based on the initial information, the analysis of separation anxiety would be the most valid. Separation anxiety is a normal toddler response when the child senses he is being sent away from those who most provide him with love and security. Crying is one way a child expresses a physical need; however, the nurse would be hasty in drawing this as a first conclusion based on what the mother has said. Nurturing overdependence or not providing structure for the toddler are inaccurate conclusions based on the information provided.

> *Nursing process step:* Analysis
> *Client needs category:* Health promotion and maintenance
> *Client needs subcategory:* Growth and development through the life span
> *Taxonomic level:* Knowledge

2. *Correct answer:* **D**

Parents can have the greatest impact on their children's self-concepts by giving the children positive feedback. Positive feedback reinforces the development of a positive self-image. Clear guidelines and structure are also important during a child's development, particularly because they give the child a sense of safety and security; however, they are not as important as positive feedback. Correction, again, is an important aspect of child rearing, but frequent correction can impair the child's development of initiative. Fluctuating limits make it difficult for a child to know his boundaries and to develop a sense of security.

> *Nursing process step:* Implementation
> *Client needs category:* Psychosocial integrity
> *Client needs subcategory:* Coping and adaptation
> *Taxonomic level:* Application

3. *Correct answer:* **A**

It's normal for a school-age child to explore genitalia, and this behavior should not be characterized as "bad." Instructing the parent to "ignore it and hope that he outgrows it" is subtly implying that the child's behavior should be of concern if it does not stop by a certain time in the future. Supervising a 6-year-old's bath will not foster the child's initiative, which is the developmental milestone for school-age children. Disciplining the child for the behavior may have an adverse affect on the child's psychosocial development.

> *Nursing process step:* Evaluating
> *Client needs category:* Health promotion and maintenance
> *Client needs subcategory:* Growth and development through the life span
> *Taxonomic level:* Application

4. *Correct answer:* **A**

Mrs. Miller is experiencing a change in her body image because of menopause. She may also be experiencing difficulty with the role change resulting from her inability to have children due to menopause. Both of these changes can affect her sense of accomplishment and her self-esteem; therefore, the self-esteem need should be the nurse's focus. The patient's physiologic, safety, and love and belonging needs, though important in planning the patient's care, are not the focus of the situation.

> *Nursing process step:* Implementation
> *Client needs category:* Health promotion and maintenance
> *Client needs subcategory:* Growth and development through the life span
> *Taxonomic level:* Application

5. *Correct answer:* **B**

Autumn is experiencing uneasiness about the curvature of her spine, which will be more evident when she wears a bathing suit. This data suggests a body image disturbance. There is no evidence of anxiety or ineffective coping. The fact that Autumn is planning to attend a swim party dispels a diagnosis of social isolation.

> *Nursing process step:* Analysis
> *Client needs category:* Psychosocial integrity
> *Client needs subcategory:* Coping and adaptation
> *Taxonomic level:* Application

6. *Correct answer:* **B**

When talking to children about death, euphemisms should be avoided. It's better to have a frank discussion recalling others who have died in the child's life. Using expressions such as "going away" can mislead the child into believing the person might return. Portraying death as sleep may provoke a sleep disturbance in the child. Telling a child that the dying person may get better is false assurance and can hamper his grieving process.

> *Nursing process step:* Implementation
> *Client needs category:* Psychosocial integrity
> *Client needs subcategory:* Coping and adaptation
> *Taxonomic level:* Application

7. *Correct answer:* **C**

Spiritual distress occurs when the individual experiences a disturbance in his personal source of strength and hope. For a child, this is most likely a parent. Changes such as moving to a new home, having another child in the family, or growing into new responsibilities can be stressful to a child, but nothing is more distressing than the loss of a parent or caretaker.

Nursing process step: Assessment
Client needs category: Psychosocial integrity
Client needs subcategory: Coping and adaptation
Taxonomic level: Application

8. *Correct answer:* **D**

A child's perception of death is influenced by a range of factors, including his previous exposure to death and what he has been taught about death. Preschoolers are most influenced by the reactions of parents and other adults to death. If the child's only exposure to death has been the death of an older person, he could perceive death from that perspective. The child could view death as punishment if that is what he has learned. The predominant perception of death by preschool-age children, however, is that death is a temporary separation. Because the child is losing someone significant and will not see that person again, it's inaccurate to infer that the death is insignificant, regardless of the child's response.

Nursing process step: Implementation
Client needs category: Psychosocial integrity
Client needs subcategory: Coping and adaptation
Taxonomic level: Application

9. *Correct answer:* **D**

Stress stimulates the adrenal cortex to increase the release of aldosterone. Aldosterone promotes the resorption of sodium, the retention of water, and the loss of potassium.

Nursing process step: Analysis
Client needs category: Physiological integrity
Client needs subcategory: Reduction of risk potential
Taxonomic level: Analysis

10. *Correct answer:* **D**

Norepinephrine is a potent vasoconstrictor that can further impair blood flow in the already compromised circulation of an individual in sickle cell crisis. Glucagon raises serum blood sugar, estrogen affects the development of female sex characteristics and reproductive function, and the antidiuretic hormone suppresses urine production.

Nursing process step: N/A
Client needs category: Physiological integrity
Client needs subcategory: Reduction of risk potential
Taxonomic level: Application

11. *Correct answer:* **D**

Allowing the patient to make as many choices as possible can reduce her sense of powerlessness. The patient-controlled device allows the patient to self-dose within prescribed parameters. Other methods of medication administration, such as intramuscular, I.V., and oral, can be administered as necessary, giving the patient a measure of control over the times of administration. The patient-controlled device, however, has the benefit of giving the patient more independence than other methods that require more of a nurse's participation.

> *Nursing process step:* Assessment
> *Client needs category:* Physiological integrity
> *Client needs subcategory:* Pharmacological and parenteral therapies
> *Taxonomic level:* Comprehension

12. *Correct answer:* **C**

In adolescent psychosocial development, the peer group is a significant mechanism of support. The absence of visits from friends while she is hospitalized may suggest that Cecily has not established a relationship with them and is therefore lacking that aspect of her normal development. Back massage is a pain-relieving therapy that is reasonable to expect from a patient experiencing Cecily's level of pain; her frequent requests do not suggest a developmental lag. Adversity with one's parents during adolescence is a normal developmental experience. Desiring privacy during hospitalization is an expected concern for all patients and should not be interpreted as evidence of a developmental lag.

> *Nursing process step:* Analysis
> *Client needs category:* Psychosocial integrity
> *Client needs subcategory:* Coping and adaptation
> *Taxonomic level:* Application

13. *Correct answer:* **C**

Self-esteem assessment is a portion of the nurse's assessment of the patient's self-concept, which also includes assessment of the patient's personal identity, body image, and role performance. The open-ended statement by the nurse, "Tell me about your sense of satisfaction with yourself," is most effective because it gives Cecily an opening for self-expression, whereas "Are you happy with yourself and your life in general?" calls for simple acknowledgment or denial, answers that will provide little insight for the nurse. Asking the patient about her likes and dislikes has more of a focus on external factors and can cause the interview to move away from the patient's self-perception. "What changes in your body most concern you?" is a question aimed at body image assessment.

Nursing process step: Assessment
Client needs category: Psychosocial integrity
Client needs subcategory: Coping and adaptation
Taxonomic level: Comprehension

14. *Correct answer:* **C**

Maslow's hierarchy of needs presents the individual's needs in ascending priority. First-level needs include oxygen, food, and shelter, which are followed by the second-level need of safety. Third-level needs are categorized as love and belonging needs, followed by the fourth level of esteem needs. The final level of needs is self-actualization.

Nursing process step: N/A
Client needs category: Psychosocial integrity
Client needs subcategory: Coping and adaptation
Taxonomic level: Knowledge

15. *Correct answer:* **B**

A nurse must ensure that the patient may pursue her convictions independent of approval from other patients. On occasion, to satisfy the needs of both patients, it may be necessary for one patient to be moved to another bed, but prohibiting a patient from discussing religion with visitors is not an appropriate nursing action. Although the physical needs of patients are often tantamount, there is no obligation for them to be the only focus of nursing care.

Nursing process step: Planning
Client needs category: Psychosocial integrity
Client needs subcategory: Coping and adaptation
Taxonomic level: Knowledge

16. *Correct answer:* **C**

In this situation in particular, the axiom "Information decreases anxiety" is most pertinent. Cecily falsely believes that a blood transfusion will lead to infection with the human immunodeficiency virus (HIV), which can lead to acquired immunodeficiency syndrome. Though there are instances in which HIV has been transmitted by way of blood transfusion, it's important for the nurse to teach Cecily that current practices have greatly reduced the risk. It's true that in severe anxiety and panic, reality can be distorted; however, in mild anxiety, reality can become more vivid. It's also true that if a myth is guiding an individual's decision making, dispelling the myth can improve the quality of the decision that is made. Concluding that the anxiety will disappear along with the myth, however, is an unreal expectation. Privacy is an important need when discussing such confidential issues as HIV infection; however, the concept that should guide the nurse in this situation is giving the patient accurate information.

Nursing process step: Planning
Client needs category: Psychosocial integrity
Client needs subcategory: Coping and adaptation
Taxonomic level: Comprehension

17. *Correct answer:* **D**

In evaluating the effect of nursing interventions, it's particularly useful to observe the behaviors that provoked the diagnosis in the first place. In Benjamin's case, evidence of his hopelessness was his refusal to participate in his own care. Therefore, a means of measuring nursing intervention success is to look at the intervention's influence on the patient's willingness to participate in his care. Shaving himself and brushing his teeth daily will show that attempts to resolve his sense of hopelessness are succeeding. The other goals listed, though perhaps useful in gauging the effectiveness of nursing care, are not as specific to Benjamin's manifestations as is the self-care goal.

Nursing process step: Planning
Client needs category: Psychosocial integrity
Client needs subcategory: Coping and adaptation
Taxonomic level: Application

18. *Correct answer:* **C**

In Orthodox Judaism, certain foods are prohibited because they are considered unclean (not kosher). Included are pork and scavengers of the sea, such as clams and crabs, which are commonly referred to as shellfish.

Nursing process step: Assessment
Client needs category: Physiological integrity
Client needs subcategory: Basic care and comfort
Taxonomic level: Comprehension

19. *Correct answer:* **D**

Physical manifestations of severe anxiety are pronounced and include tremors, sweating, and nausea, as well as urinary frequency and palpitations. Elevated heart rate and respiratory rate are evident in all forms of anxiety and are not specifically diagnostic of severe anxiety. Choking and chest pain are most associated with panic. Pupil dilation and dry mouth, though perhaps present in severe anxiety, are more associated with a diagnosis of moderate anxiety.

Nursing process step: Assessment
Client needs category: Physiological integrity
Client needs subcategory: Reduction of risk potential
Taxonomic level: Knowledge

20. *Correct answer:* **D**

When severely anxious, the individual senses a loss of control; stimuli, such as noise and activity, can heighten his anxiety. Reducing the demands placed on him and lessening the distracting stimuli in his environment can help lower his level of anxiety. Severe anxiety impairs decision making and the ability to follow instructions; therefore, the nurse should avoid asking him to make choices or answer questions. Instructing the individual to be still can place unnecessary demands on him and prohibit him from using activity as a coping mechanism for anxiety.

> *Nursing process step:* Implementation
> *Client needs category:* Psychosocial integrity
> *Client needs subcategory:* Coping and adaptation
> *Taxonomic level:* Application

21. *Correct answer:* **C**

Lorazepam (Ativan) is a benzodiazepine, the most widely used class of antianxiety drugs (anxiolytics). Diphenhydramine (Benadryl), a histamine antagonist (antihistamine), has sedative effects, but it isn't specifically for the treatment of anxiety. Flumazenil (Romazicon) is a benzodiazepine antagonist that counteracts the effect of an antianxiety agent. Methylphenidate (Ritalin) is a central nervous system stimulant that would more likely heighten anxiety in the individual.

> *Nursing process step:* N/A
> *Client needs category:* Physiological integrity
> *Client needs subcategory:* Pharmacological and parenteral therapies
> *Taxonomic level:* Application

22. *Correct answer:* **D**

Progressive muscle relaxation is a method of stress management in which the individual tenses and relaxes various muscle groups. The patient is taught to stop the technique and massage the affected muscle group if cramping occurs. Furthermore, the patient is taught to close his eyes during the exercise, to avoid muscle tightening to the point of pain, and to breathe slowly and deeply during the relaxation technique.

> *Nursing process step:* Evaluation
> *Client needs category:* Psychosocial integrity
> *Client needs subcategory:* Coping and adaptation
> *Taxonomic level:* Application

23. *Correct answer:* **D**

According to Erikson's stages of psychosocial development, each stage of life has a conflict with which it's associated. Early adulthood involves the conflict of intimacy versus isolation. Adolescence is characterized by the conflict of identity versus role confusion. Industry versus inferiority is the conflict of the school-age child. Integrity versus despair is the conflict of later adulthood.

> *Nursing process step:* N/A
> *Client needs category:* Health promotion and maintenance
> *Client needs subcategory:* Growth and development through the life span
> *Taxonomic level:* Application

24. *Correct answer:* **B**

Encouraging a patient to be as independent as possible will promote self-reliance and self-confidence, both of which are components of a healthy self-esteem. An effective means of encouraging this independence is enlisting the patient in the planning of his own care. Teaching self-care measures and eliciting his peers to encourage him will be beneficial, but they are not as focused on fostering a healthy self-esteem. Moving Mr. Martin to a different hospital environment isn't indicated based on the data given.

> *Nursing process step:* Planning
> *Client needs category:* Psychosocial integrity
> *Client needs subcategory:* Coping and adaptation
> *Taxonomic level:* Application

25. *Correct answer:* **D**

Mr. Martin is going through the grief-loss process because he has permanently lost the use of his legs due to paralysis. His comment suggests that he's in denial, the first stage of the process. Denial is a coping mechanism that allows the individual time to assimilate the major changes associated with body function loss. Anger is the phase of the process that follows denial, bargaining is the third phase, and acceptance is the final phase.

> *Nursing process step:* Assessment
> *Client needs category:* Psychosocial integrity
> *Client needs subcategory:* Coping and adaptation
> *Taxonomic level:* Application

26. *Correct answer:* **D**

The patient is defensive about being seen holding a male friend's hand and expects others to misinterpret or misjudge the act. It's most important for the nurse to reflect understanding, empathy, objectivity, and a nonjudgmental attitude toward the patient. Although the patient may be experiencing an identity disturbance, the situation doesn't clearly suggest this implication. It's important for nurses to foster healthy self-esteem in their patients; however, this is not the main concept that should guide the nurse in this situation. Accentuating the positive by conducting one's practice with a nonjudgmental attitude is a more genuine and effective concept than disguising the negative feelings the nurse may have toward the patient's behavior.

> *Nursing process step:* Implementation
> *Client needs category:* Psychosocial integrity
> *Client needs subcategory:* Coping and adaptation
> *Taxonomic level:* Comprehension

27. *Correct answer:* **B**

Regardless of the nurse's personal spiritual beliefs or feelings, it's important for her to meet the patient's expressed need. An effective way to do that in this situation is to let the patient take the lead. Telling the patient to pray or have a moment of silence blocks communication and removes the patient from the decision making. Calling the chaplain counters the patient's request for the nurse to pray with him and can diminish the trust the patient has in the nurse.

> *Nursing process step:* Implementation
> *Client needs category:* Psychosocial integrity
> *Client needs subcategory:* Coping and adaptation
> *Taxonomic level:* Application

28. *Correct answer:* **D**

Stress reduces the protective mechanisms in the upper GI tract, including a reduction in gastric mucosa production. Stress increases the secretion of substances that make the gastric juices acidic, which lowers the gastric pH (more acid). Rather than being blocked, the parietal cells actually become more active during stress. Stress increases stimulation of the vagus nerve, which in turn causes an increase in stomach acid.

> *Nursing process step:* N/A
> *Client needs category:* Physiological integrity
> *Client needs subcategory:* Pharmacological and parenteral therapies
> *Taxonomic level:* Comprehension

29. *Correct answer:* **C**

Ranitidine (Zantac) is a histamine antagonist, ciprofloxacin (Cipro) is an anti-infective, morphine sulfate (Morphine) is a narcotic analgesic, and sucralfate (Carafate) is an antiulcer agent.

> *Nursing process step:* N/A
> *Client needs category:* Physiological integrity
> *Client needs subcategory:* Pharmacological and parenteral therapies
> *Taxonomic level:* Knowledge

30. *Correct answer:* C

Early grief involves shock, disbelief, and denial; therefore, statements such as "it's a dream" and "it didn't really happen" are expected reactions in that stage. Job retraining questions are more suggestive of either the acceptance phase or dysfunctional grief, in which the individual is failing to grieve. Anger is the second phase of grief. Isolation is more suggestive of the depression phase of grief.

> *Nursing process step:* Analysis
> *Client needs category:* Psychosocial integrity
> *Client needs subcategory:* Coping and adaptation
> *Taxonomic level:* Comprehension

31. *Correct answer:* C

Extreme behaviors such as Ms. Livingston's distorted response to the ambulance are indicative of panic. In the lesser degrees of anxiety, the patient's typical behavior is changed, but not to the exaggerated level that is seen in panic. Mild anxiety is characterized by increased alertness and enhanced focus. Moderate anxiety is most characterized by escalated body functions, including such behaviors as pacing and accelerated speech. Severe anxiety involves a generalized bodily response, which includes such reactions as urinary urgency, nausea, dizziness, and numbness and tingling sensations.

> *Nursing process step:* Analysis
> *Client needs category:* Psychosocial integrity
> *Client needs subcategory:* Psychosocial adaptation
> *Taxonomic level:* Application

32. *Correct answer:* D

When a patient is experiencing panic, it's most important for the nurse to remain with the patient to provide physical and verbal assurance as well as to protect her from further injury. During panic, teaching Ms. Livingston about effective coping mechanisms and encouraging her to discuss her feelings are less appropriate interventions because they can agitate the patient even more. Use of restraints requires a doctor's order and can cause injury to the skin and joints.

> *Nursing process step:* Planning
> *Client needs category:* Psychosocial integrity
> *Client needs subcategory:* Coping and adaptation
> *Taxonomic level:* Application

33. *Correct answer:* B

A number of measures are available to distract and relax an individual, allowing her to at least partly forget about her worries and be clearer in thought and communication. Among these measures are massage and heat. The patient's calmness and clearer communication were most likely a result of the hydrotherapy experience, not a result of the environmental change or the act of being submerged in water. Hydrotherapy does not physically numb the patient.

Nursing process step: Analysis
Client needs category: Physiological integrity
Client needs subcategory: Basic care and comfort
Taxonomic level: Application

34. *Correct answer:* **B**

Spiritual distress occurs when the individual's belief system is threatened. Manifestations include a sense of guilt or of being punished. Isolated comments, such as wanting a secular burial for loved ones, discussing one's religious history, and expressing sadness for one's loss, do not indicate spiritual distress in and of themselves.

Nursing process step: Assessment
Client needs category: Psychosocial integrity
Client needs subcategory: Coping and adaptation
Taxonomic level: Comprehension

35. *Correct answer:* **C**

Option C demonstrates projection, a defense mechanism in which the individual ascribes responsibility to others. Option A is rationalization, option B is displacement, and option D is denial.

Nursing process step: Analysis
Client needs category: Psychosocial integrity
Client needs subcategory: Psychosocial adaptation
Taxonomic level: Application

36. *Correct answer:* **A**

Displacement exists when the individual focuses his true response on a more vulnerable or safer target than the target for which it was intended, such as a man yelling at his wife after his boss reprimands him. Avoiding responsibility for one's behavior is denial. Behaving in a less mature fashion is regression. Fabricating plausible reasons for unacceptable behavior is rationalization.

Nursing process step: Implementation
Client needs category: Psychosocial integrity
Client needs subcategory: Coping and adaptation
Taxonomic level: Comprehension

37. *Correct answer:* **B**

Fostering healthy self-concepts in children requires a range of parental behaviors, including providing structure for the children. Structure gives children security and the parameters for success. Sincere and timely praise is essential in fostering self-concept in children — withholding praise can discourage children. On occasion, it may be necessary to supervise children's interaction with friends, but this isn't a primary principle in fostering healthy self-concepts. Free reign in self-expression does not teach a child to live in a society of organization and limitations.

> *Nursing process step:* Implementation
> *Client needs category:* Psychosocial integrity
> *Client needs subcategory:* Coping and adaptation
> *Taxonomic level:* Application

38. *Correct answer:* **D**

A class on accurate medication usage meets the safety needs of clients; therefore, it would be planned first when using Maslow's hierarchy of needs as a guide. Organizing volunteers focuses on self-actualization needs, which are the highest priority in the hierarchy. Potlucks, dances, and singles mixers focus on love and belonging needs, which are the third priority in the hierarchy. The speakers' bureau would focus on esteem needs, the fourth level of the hierarchy.

> *Nursing process step:* Planning
> *Client needs category:* Physiological integrity
> *Client needs subcategory:* Basic care and comfort
> *Taxonomic level:* Application

39. *Correct answer:* **B**

Three spiritual needs underlying all religions and common to all people are (1) a life with meaning and purpose, (2) a need for love and relatedness, and (3) a need for forgiveness.

> *Nursing process step:* Planning
> *Client needs category:* Psychosocial integrity
> *Client needs subcategory:* Coping and adaptation
> *Taxonomic level:* Comprehension

40. *Correct answer:* **B**

Stress has been shown to diminish the body's immune response by causing the increased release of cortisol, an anti-inflammatory hormone. For example, research has demonstrated that stress increases one's risk for the common cold. Stress isn't always deleterious — in fact, it's useful in the fight-or-flight response. Stress increases metabolism, one's use of energy (kilocalories). Because stress stimulates the sympathetic autonomic nervous system, it decreases secretion by digestive glands and slows gastrointestinal motility, thereby lessening digestion.

> *Nursing process step:* Planning
> *Client needs category:* Physiological integrity
> *Client needs subcategory:* Reduction of risk potential
> *Taxonomic level:* Application

41. *Correct answer:* **A**

Self-actualization has a focus on respecting, showing affection toward, and helping a person outside of himself. It's the highest achievement in Maslow's hierarchy of basic human needs. While being able to see one's heritage as successful, to view oneself as good, and to take on new challenges are positive individual outlooks, they do not point to self-actualization as accurately as taking interests outside oneself.

> *Nursing process step:* Evaluation
> *Client needs category:* Health promotion and maintenance
> *Client needs subcategory:* Growth and development through the life span
> *Taxonomic level:* Comprehension

42. *Correct answer:* **C**

Using the therapeutic communication technique of reflection, the nurse encourages the patient to express his concerns about his sexuality. This can begin the problem-solving process for the patient. Response A demonstrates empathy but postpones the topic the patient wants to talk about. Response B is judgmental in nature and blocks communication between the patient and nurse. Response D may benefit the patient, but it's premature and would best occur after the patient has had an opportunity to express his concerns.

> *Nursing process step:* Assessment
> *Client needs category:* Psychosocial integrity
> *Client needs subcategory:* Coping and adaptation
> *Taxonomic level:* Application

43. *Correct answer:* **D**

Endorphins are narcotic-like substances that lock into the narcotic receptors at nerve endings in the brain and spinal cord and block the transmission of pain signals.

> *Nursing process step:* Implementation
> *Client needs category:* Psychosocial integrity
> *Client needs subcategory:* Coping and adaptation
> *Taxonomic level:* Knowledge

44. *Correct answer:* **C**

Endorphin levels are depleted by recurrent stress, prolonged pain, and the pro-longed use of narcotics and alcohol. Endorphin levels are increased during brief periods of pain or stress, physical exercise, sexual activity, trauma, some types of acupuncture, and transcutaneous nerve stimulation.

> *Nursing process step:* N/A
> *Client needs category:* Psychosocial integrity
> *Client needs subcategory:* Coping and adaptation
> *Taxonomic level:* Comprehension

45. *Correct answer:* **B**

Moderate anxiety stimulates the sympathetic autonomic nervous system, leading to increased metabolism (sometimes seen in a slight body temperature elevation) and elevation in the individual's heart rate, respiratory rate, and blood pressure.

> *Nursing process step:* Analysis
> *Client needs category:* Psychosocial integrity
> *Client needs subcategory:* Coping and adaptation
> *Taxonomic level:* Knowledge

46. *Correct answer:* **C**

Stress stimulates an increase in the body's release of catecholamines and cortisol. Both of these hormones cause an increase in serum glucose levels by mobilizing glucose reserves and lowering endogenous insulin release. Although stress does cause peripheral vasoconstriction, it does not alter the accuracy of fingerstick blood values. Glucagon and insulin function in a reciprocal manner, so concur-rent elevation in their levels is unlikely.

> *Nursing process step:* Implementation
> *Client needs category:* Psychosocial integrity
> *Client needs subcategory:* Coping and adaptation
> *Taxonomic level:* Comprehension

47. *Correct answer:* **D**

Mild anxiety makes an individual's perception more focused and acute. It height-ens one's awareness, problem-solving skills, and attention to details. Rather than promoting energy storage by the body, mild anxiety activates metabolism and causes the body to use stored energy. Anxiety in any form is more likely to cause insomnia than to promote sleep. Although mild anxiety may improve one's use of the organs of sight and sound, it does not improve their function.

Nursing process step: Implementation
Client needs category: Psychosocial integrity
Client needs subcategory: Coping and adaptation
Taxonomic level: Knowledge

48. *Correct answer:* **A**

An individual experiencing such a loss as loss of vision is expected to gain the support of those close to him and of his caregivers, such as nurses. By capitalizing on his support system, the patient is demonstrating achievement of this goal. Expressing certainty of regaining his sight and setting a date to return to his previous job most likely demonstrate that he is in the denial phase of grief and loss. The purchases, at best, are only superficial evidence of the patient accepting his blindness.

Nursing process step: Evaluation
Client needs category: Psychosocial integrity
Client needs subcategory: Coping and adaptation
Taxonomic level: Analysis

49. *Correct answer:* **B**

Impotence is defined as the inability to achieve penile erection. It also involves being unable to maintain an erection to the point of ejaculation. Infertility is the term used when a man's semen is incapable of impregnating a woman. Disinterest in sexual intimacy can be labeled with a number of terms including celibacy and asexuality.

Nursing process step: Evaluation
Client needs category: Psychosocial integrity
Client needs subcategory: Coping and adaptation
Taxonomic level: Application

50. *Correct answer:* **B**

Stress in the workplace is most likely experienced by individuals with less autonomy and control. Independence in decision making allows a person to experience self-actualization, which modifies stress. Deadlines and alternate schedules are not as stressful as is the lack of control and independence in the workplace.

Nursing process step: Planning
Client needs category: Psychosocial integrity
Client needs subcategory: Coping and adaptation
Taxonomic level: Comprehension

51. *Correct answer:* **D**

Expressing her knowledge of the alphabet and numbers identifies Keyshaun's sense of competence, which is evidence of a positive self-concept in children. Submitting, paying attention, and being the recipient of high praise may suggest that she is in an environment that nurtures a positive self-concept, but they do not display its existence as well as her statement of accomplishment.

> *Nursing process step:* N/A
> *Client needs category:* Psychosocial integrity
> *Client needs subcategory:* Coping and adaptation
> *Taxonomic level:* Application

52. *Correct answer:* **A**

Keyshaun's mother has expressed consternation about her daughter receiving the immunizations, and it's most appropriate for the nurse to allow her to express her thoughts. Asking the mother, "Would you like to talk about it?" opens the way for the mother to ventilate. Responses B and C can be interpreted as dismissive and argumentative, affecting the patient-nurse rapport. Response D is inappropriate because it ignores the mother's concerns.

> *Nursing process step:* Assessment
> *Client needs category:* Psychosocial integrity
> *Client needs subcategory:* Coping and adaptation
> *Taxonomic level:* Application

53. *Correct answer:* **B**

The developmental task for preschoolers is initiative. Identity is associated with adolescence, intimacy with young adults, and integrity with later adulthood.

> *Nursing process step:* Assessment
> *Client needs category:* Health promotion and maintenance
> *Client needs subcategory:* Growth and development through the life span
> *Taxonomic level:* Comprehension

54. *Correct answer:* **A**

Empathy, the understanding of another's perception of a situation, is key to establishing a therapeutic rapport with a patient. Empathy encourages a patient's trust and promotes the patient's self-expression. In this situation, it's the most important aspect of the nursing role that should be expressed.

> *Nursing process step:* Planning
> *Client needs category:* Psychosocial integrity
> *Client needs subcategory:* Coping and adaptation
> *Taxonomic level:* Application

55. *Correct answer:* **A**

The Roman Catholic practice is to baptize infants soon after birth, especially if they are ill and hospitalized. Circumcision may be performed on the infant of a Roman Catholic, but it isn't a considered a Roman Catholic practice. Last rites or sacraments of the sick wouldn't be appropriate for the mother of the child.

> *Nursing process step:* Planning
> *Client needs category:* Health promotion and maintenance
> *Client needs subcategory:* Growth and development through the life span
> *Taxonomic level:* Comprehension

56. *Correct answer:* **A**

Repression involves dismissing memories from one's consciousness. Despite her boyfriend's suicide, Ms. Alba is stating that he'll help her care for the infant—this is repression.

> *Nursing process step:* Assessment
> *Client needs category:* Psychosocial integrity
> *Client needs subcategory:* Coping and adaptation
> *Taxonomic level:* Application

57. *Correct answer:* **C**

Trust is the developmental landmark of infancy and can be nurtured by meeting the needs of the infant with care, concern, warmth, and consistency. The infant's need for sleep, food, and warmth is irrefutable, but it cannot overshadow his need for touch and attentive concern.

> *Nursing process step:* Implementation
> *Client needs category:* Health promotion and maintenance
> *Client needs subcategory:* Growth and development through the life span
> *Taxonomic level:* Comprehension

58. *Correct answer:* **A**

Assisting a patient past a subject of denial can be achieved by therapeutically identifying areas of his denial. Of course, this assistance requires that a nurse have rapport and trust with the patient. Literature has the possibility of being most effective when used in conjunction with nurse-patient interaction. The referral can be useful, but it shouldn't supersede the nurse's direct intervention with the patient. Because Mr. Denton has a sexually transmitted disease, there is a public health concern; encouraging his denial can put others at risk.

> *Nursing process step:* Evaluation
> *Client needs category:* Psychosocial integrity
> *Client needs subcategory:* Coping and adaptation
> *Taxonomic level:* Knowledge

59. *Correct answer:* **C**

Bargaining involves deal making by the patient and is the third phase in the five stages of dying. Anger is the second stage, depression is the fourth stage, and acceptance is the last stage.

> *Nursing process step:* Assessment
> *Client needs category:* Psychosocial integrity
> *Client needs subcategory:* Coping and adaptation
> *Taxonomic level:* Knowledge

60. *Correct answer:* **B**

The patient's unwillingness to look at himself in the mirror is most indicative of body image disturbance. All of the other nursing diagnoses mentioned may exist in the individual, but they require more supporting data.

> *Nursing process step:* Analysis
> *Client needs category:* Psychosocial integrity
> *Client needs subcategory:* Coping and adaptation
> *Taxonomic level:* Knowledge

61. *Correct answer:* **B**

In intervening in the patient's body image disturbance, it's important for the nurse to earn the patient's trust and develop a rapport. Because the patient has asked the nurse to shave him, it's an ideal opportunity for the nurse to demonstrate regard, which will bolster the patient's trust in the nurse. Instructing the patient to shave himself overlooks his consternation about the way he looks. Suggesting he postpone shaving ignores his request, and presenting him with the cliché, "accept yourself before others can accept you" isn't therapeutic.

> *Nursing process step:* Implementation
> *Client needs category:* Psychosocial integrity
> *Client needs subcategory:* Coping and adaptation
> *Taxonomic level:* Application

62. *Correct answer:* **A**

In this case, the patient's sexuality carries with it the risk of communicating disease-producing organisms. This risk is important for the nurse to communicate because of concern for both individuals and the community. Although sexuality may not be addressed in every patient care situation, it should not be avoided until a patient brings it up or a sexuality problem develops. Sexual practices, like many practices, do exist on a continuum, but this information is immaterial in this situation.

Nursing process step: Planning
Client needs category: Psychosocial integrity
Client needs subcategory: Coping and adaptation
Taxonomic level: Application

63. *Correct answer:* **B**

Morphine sulfate (Morphine) is a narcotic analgesic that has both pain-relieving
and anxiety-reducing properties. Diazepam (Valium), phenobarbital (Luminal),
and prochlorperazine (Compazine) all have sedative or anxiety-reducing capabilities, but they're not painkillers.

Nursing process step: N/A
Client needs category: Physiological integrity
Client needs subcategory: Pharmacological and parenteral therapies
Taxonomic level: Knowledge

64. *Correct answer:* **B**

The most common adverse effect of anxiolytics is drowsiness and confusion.
Because of the central nervous system depression caused by anxiolytics, a lowering of the blood pressure and respiratory rate can occur. Cardiac arrhythmias
may result from hypoxemia caused by respiratory depression, but they aren't a
common adverse effect of anxiolytics.

Nursing process step: N/A
Client needs category: Physiological integrity
Client needs subcategory: Pharmacological and parenteral therapies
Taxonomic level: Knowledge

65. *Correct answer:* **B**

Fatherhood isn't limited to men who are fertile, and it can be accomplished in
other ways. Option B is both frank and encouraging. Option A is true, but it
doesn't portray the warmth necessary in this situation. Option C seems cheery,
but it has a tone of false reassurance. Option D has an abrasive tone and does not
answer the patient's question.

Nursing process step: Implementation
Client needs category: Psychosocial integrity
Client needs subcategory: Coping and adaptation
Taxonomic level: Application

66. *Correct answer:* **D**

A body image change can affect the individual's relationships, including a
diminution in intimacy. His wife's report that there has been no impairment in
their intimacy suggests that the patient is coping with his body image change. His

intentions to begin exercising and his assertion that he's the same as he has always been are not as specific to the body image concern. Preferring not to talk about his surgery may suggest denial of the change in his body rather than coping.

Nursing process step: Evaluation
Client needs category: Psychosocial integrity
Client needs subcategory: Coping and adaptation
Taxonomic level: Application

67. *Correct answer:* **B**

The primary neurohormone in the mobilization of sexual behavior is dopamine. An androgen is from a class of steroids secreted by the adrenal cortex; this class may have some influence in libido, but they aren't the main mobilizer in sexual behavior. Serotonin inhibits sexual behavior. Testosterone is responsible for the development of secondary sexual characteristics and isn't a neurohormone.

Nursing process step: Implementation
Client needs category: Physiological integrity
Client needs subcategory: Basic care and comfort
Taxonomic level: Knowledge

68. *Correct answer:* **D**

A tenet of Protestantism is that life events are the result of God's will. Although Protestants mostly agree that there are consequences in life for certain behaviors, punishment is an event that occurs after one's life is complete. Chance, fate, and lack of faith are not reasons Protestants generally give for acquiring an illness.

Nursing process step: Evaluation
Client needs category: Psychosocial integrity
Client needs subcategory: Coping and adaptation
Taxonomic level: Analysis

69. *Correct answer:* **D**

Role playing is an active simulation in which the nurse plays the role of the person with whom the patient will be interacting. Coaching, presenting, and acting out are techniques that place the patient in a passive role, whereas role playing actually engages the patient in the problem-solving process.

Nursing process step: Implementation
Client needs category: Psychosocial integrity
Client needs subcategory: Coping and adaptation
Taxonomic level: Application

70. *Correct answer:* **C**

The major psychosocial developmental tasks of younger adulthood have to do with vocation and relationship choices. The nurse's open-ended statement can result in useful information about the patient's work and relationships. Having the patient reflect on past life experiences does not address the issue of the individual's current development. The question about starting a family assumes that this is an expectation of adulthood; however, many adults don't desire to be parents.

> *Nursing process step:* Assessment
> *Client needs category:* Health promotion and maintenance
> *Client needs subcategory:* Growth and development through the life span
> *Taxonomic level:* Comprehension

71. *Correct answer:* **C**

The developmental landmark for this patient's age-group is intimacy; having difficulty maintaining friendships suggests a lag in achieving this landmark. Changing jobs every 2 years may suggest upward mobility or a desire for change; abandoning long-held beliefs isn't unusual in this age-group, and neither is worrying about not having children.

> *Nursing process step:* Assessment
> *Client needs category:* Psychosocial integrity
> *Client needs subcategory:* Coping and adaptation
> *Taxonomic level:* Comprehension

72. *Correct answer:* **D**

Black Muslims stress the importance of cooperation among black business and education in the elevation of the self-esteem of its adherents. Prophecy is a focus of such religions as Seventh-Day Adventism and Fundamental Baptist. Reincarnation is associated with Far Eastern religions, and celibacy is a practice in a number of the world's religions.

> *Nursing process step:* Evaluation
> *Client needs category:* Psychosocial integrity
> *Client needs subcategory:* Coping and adaptation
> *Taxonomic level:* Comprehension

73. *Correct answer:* **A**

Particularly in isolation, a patient needs nursing to advocate family involvement in the care. This family involvement is especially important in meeting the love and belonging needs identified by Maslow's hierarchy of needs. None of the other measures are more important than, or are substitutes for, family involvement.

Nursing process step: Implementation
Client needs category: Psychosocial integrity
Client needs subcategory: Coping and adaptation
Taxonomic level: Application

74. *Correct answer:* **D**

The meeting of sexual needs varies from one individual to another; however, a common aspect of meeting this need is providing privacy. The other interventions are less focused on the specific need she may have to express her sexual needs.

Nursing process step: Planning
Client needs category: Health promotion and maintenance
Client needs subcategory: Growth and development through the life span
Taxonomic level: Application

75. *Correct answer:* **C**

One of the "gifts of the spirit" practiced by some of the charismatic faiths is the laying on of hands, a practice that is used in dealing with illness. The other answers are not associated with charismatic faiths.

Nursing process step: Assessment
Client needs category: Psychosocial integrity
Client needs subcategory: Coping and adaptation
Taxonomic level: Knowledge

76. *Correct answer:* **C**

A milestone of middle adulthood involves fulfilling civic and social responsibilities. Addressing unresolved conflicts is a life-span ambition not specifically identified with middle adulthood. Reconsidering career choices may be both necessary and growth-producing, but it isn't a milestone. Youthfulness is widely desired, but this desire may be a reflection of a weakness in coping or even of a self-image disturbance.

Nursing process step: N/A
Client needs category: Health promotion and maintenance
Client needs subcategory: Growth and development through the life span
Taxonomic level: Application

77. *Correct answer:* **D**

Erikson's developmental stage for Ms. Brown is generativity versus stagnation. Stagnation depicts failure to meet the developmental milestone of this stage. Despair is associated with later adulthood, isolation with young adulthood, and role confusion with adolescence.

Nursing process step: N/A
Client needs category: Health promotion and maintenance
Client needs subcategory: Growth and development through the life span
Taxonomic level: Knowledge

78. *Correct answer:* **D**

Stress and pain stimulate a sympathetic autonomic nervous system response. A result of this response is relaxation of the urinary bladder and contraction of the urethral sphincter. Sympathetic stimulation increases pupil dilation (mydriasis), bodily oxygen demands, and bronchial lumen size (bronchodilation).

Nursing process step: Analysis
Client needs category: Physiological integrity
Client needs subcategory: Reduction of risk potential
Taxonomic level: Analysis

79. *Correct answer:* **B**

The white undergarment worn by some Mormons reminds the individual of the sacred covenants she has made with God. The nurse should understand that the garment is to be worn at all times. Asking the individual about the garment, seeing it, and touching it are not prohibited.

Nursing process step: Planning
Client needs category: Psychosocial integrity
Client needs subcategory: Coping and adaptation
Taxonomic level: Application

80. *Correct answer:* **A**

Hopelessness is characterized by an inability to muster enough energy to meet even basic needs. Atheism is a philosophical preference, not necessarily a depiction of one's level of hope. Anxiety or fear are expected responses to delays in receiving a organ transplant. Infrequent visits from friends and family may cause hopelessness, but they aren't evidence of it.

Nursing process step: Assessment
Client needs category: Psychosocial integrity
Client needs subcategory: Psychosocial adaptation
Taxonomic level: Application

81. *Correct answer:* **B**

An individual's role performance or expectations involve such concerns as occupation, family life, and use of time; however, the most important information to gather is the individual's own perception of his role.

Nursing process step: Assessment
Client needs category: Psychosocial integrity
Client needs subcategory: Coping and adaptation
Taxonomic level: Analysis

82. *Correct answer:* **C**

It's important for a nurse to demonstrate acceptance of a patient, even if she disagrees with the ideas the patient communicates. This demonstration is best accomplished by active listening. A block to nurse-patient communication is changing the subject, which occurs in response A. Response B is a judgmental comment, which is another block to communication. Response D is dishonest in that the nurse who disagrees with the patient's point of view tells him she respects it. A nurse should respect a patient's *right* to express a point of view, but she is under no obligation to respect the point of view expressed.

Nursing process step: Implementation
Client needs category: Psychosocial integrity
Client needs subcategory: Coping and adaptation
Taxonomic level: Application

83. *Correct answer:* **D**

In helping grieving family members, it's important for the nurse to realize the concept that needs met by key people in our lives can be met in other ways and by other people. Denial is an expected phase in the grief process; however, it isn't therapeutic nursing to promote the family's denial. Grief work begins early in the death process, not after death. Nursing is most effective when empathy is demonstrated. Sympathy is less objective than empathy and can impair the nurse's effectiveness with the family.

Nursing process step: Planning
Client needs category: Psychosocial integrity
Client needs subcategory: Coping and adaptation
Taxonomic level: Application

84. *Correct answer:* **B**

The most important priority in caring for a dying patient is the communication of caring. The other nursing attributes are necessary in effective nursing care, but none are as important as caring when meeting the needs of a dying patient.

Nursing process step: Implementation
Client needs category: Psychosocial integrity
Client needs subcategory: Coping and adaptation
Taxonomic level: Application

85. *Correct answer:* **C**

Role conflict occurs in the individual when the expectations of one of the individual's roles conflicts or interferes with the performance of another of the individual's roles. An example of this conflict is the case of a parent who runs a business in which another family member is an employee. The roles that may conflict in this case would be parent and employer.

Nursing process step: N/A
Client needs category: Psychosocial integrity
Client needs subcategory: Coping and adaptation
Taxonomic level: Knowledge

86. *Correct answer:* **D**

The patient is giving the nurse a cue that he is concerned about his death. An effective way of encouraging his further communication about the issue is to use a reflective comment. Reflection is stating what the nurse perceives the patient to be feeling. The other responses do not address the patient's concern, ignore it, or attempt to pass it on to another professional.

Nursing process step: Implementation
Client needs category: Psychosocial integrity
Client needs subcategory: Coping and adaptation
Taxonomic level: Application

CHAPTER 7

PATIENTS WITH SPECIAL NEEDS

QUESTIONS

1. During a neurosensory assessment, the nurse observes that a patient has a slightly dilated pupil in his right eye. The nurse understands that this can be explained by nonconduction of the:

    A. second cranial nerve (optic).
    B. third cranial nerve (oculomotor).
    C. fourth cranial nerve (trochlear).
    D. seventh cranial nerve (facial).

2. Immediately following a motor vehicle accident, a teenager is diagnosed as a quadriplegic. She's placed on a tilt table for half an hour while the head of the table is elevated to a 30-degree angle. Each day the angle is increased, and the time spent on the table is also increased. The nurse explains to the patient that this procedure is intended to:

    A. improve circulation to the extremities.
    B. prevent hypotension.
    C. prevent loss of calcium from the bones.
    D. prevent pressure sores.

3. A 35-year-old male patient complaining of chronic progressive chorea and mental deterioration is admitted to the unit. The nurse recognizes that these characteristics indicate a disease that results in degeneration of the basal ganglia and cerebral cortex. The disease is called:

    A. Guillain-Barré syndrome.
    B. Huntington's disease.
    C. multiple sclerosis (MS).
    D. myasthenia gravis.

4. In caring for a patient recovering from an acute head injury, the nurse understands that the onset of seizures most likely will occur:

    A. 6 months to 2 years after the head injury.
    B. 3 to 5 years after the head injury.
    C. within the first 6 months after the head injury.
    D. within the first weeks posttrauma.

5. A patient is admitted to the rehabilitation unit with a diagnosis of quadriplegia. Planning nursing care for a quadriplegic patient is based on the nurse's knowledge that quadriplegia results from injuries to what level of the spinal cord?

    A. Cervical spine levels C1 to C4
    B. Cervical spine levels C1 to C8
    C. Cervicothoracic spine levels C3 to T4
    D. Thoracolumbar spine levels T1 to L4

6. A patient is diagnosed with an upper motor neuron lesion. The nurse expects that the patient will experience which of the following conditions?

    A. Excessive parasympathetic stimulation
    B. Flaccid paralysis
    C. Spastic paralysis
    D. Temporary paralysis

7. The nurse understands that normal physiological changes in the eye associated with aging include:

    A. decreased elasticity of the lens and a slower pupillary light reflex.
    B. decreased strabismus and decreased tearing.
    C. increased elasticity of the lens and a slower pupillary light reflex.
    D. increased strabismus and decreased tearing.

8. The nurse understands that physical changes associated with aging include:

    A. decreased physical strength and endurance.
    B. increased muscular coordination.
    C. increased tolerance to cold.
    D. a tendency to gain weight.

9. Psychogenic pain is best defined as pain:

    A. due to a lesion in the central nervous system.
    B. due to emotional factors without anatomic or physiologic explanation.
    C. originating in the internal organs.
    D. due to an amputation.

10. When preparing a nursing plan of care for an elderly patient, the nurse understands that major fears of elderly people include:

    A. fear of acute illness.
    B. fear of economic dependency.
    C. fear of physical and economic dependency.
    D. fear of physical dependency.

SITUATION: *Alan Simpson has suffered a cerebrovascular accident in the left hemisphere of his brain.*

Questions 11 to 14 refer to this situation.

11. The nurse anticipates that Mr. Simpson will have paresis on the:

    A. four extremities of his body.
    B. left side of his body.
    C. right side of his body.
    D. right side of his body and the left side of his face.

12. Mr. Simpson has difficulty speaking, but he can communicate through gestures. The term that best describes this condition is:

    A. acephalia.
    B. accoucheur.
    C. akathisia.
    D. aphasia.

13. As part of the long-term planning for Mr. Simpson, the nurse should:

    A. begin associating words with physical objects.
    B. ask the patient questions and wait as long as it takes for him to verbally respond.
    C. consult with the patient's doctor concerning a speech therapy program.
    D. help the patient accept this disability as permanent.

14. The action most helpful in communicating with Mr. Simpson is:

    A. conversing in a rapid manner.
    B. giving directions on a two-at-a-time basis.
    C. speaking in a louder-than-usual tone of voice.
    D. using gestures to accompany the spoken word.

SITUATION: *Kirk Richards is admitted to the hospital with a disease characterized by the rapid development of symmetrical weakness and lower motor neuron flaccid paralysis that ascends to the upper extremities.*

Questions 15 to 17 refer to this situation.

15. The nurse recognizes that Mr. Richards's symptoms are characteristics of which of the following diseases?

A. Guillain-Barré syndrome
B. Huntington's disease
C. Multiple sclerosis (MS)
D. Myasthenia gravis

16. The nurse anticipates that the doctor will order which of the following treatments for Mr. Richards?

A. High-dose steroid therapy
B. I.V. wide-spectrum antibiotic therapy
C. The monitoring of respiratory vital capacity every 2 hours
D. Pyridostigmine bromide (Mestinon)

17. Mr. Richards deteriorates, suffering respiratory failure and requiring transfer to the intensive care unit for continuing care. Based on her knowledge of the disease, the nurse knows which of the following facts about Mr. Richards's prognosis?

A. Death usually occurs in 1 to 2 months.
B. Prognosis is poor or variable and may progress rapidly or slowly.
C. Recovery is complete and spontaneous in about 95% of all patients.
D. The patient will most likely have severe residual lower extremity weakness requiring long-term supportive care.

SITUATION: *Walter Tyson is a 67-year-old man who's admitted with a diagnosis of a brain aneurysm.*

Questions 18 to 20 refer to this situation.

18. Based on the diagnosis of a brain aneurysm, the nurse anticipates which of the following orders from the doctor?

A. Ambulate the patient the length of the hallway three times a day.
B. Administer hydralazine, prednisone, and diazepam around the clock.
C. Administer I.V. phenytoin and phenobarbitol.
D. Administer subcutaneous terbutaline alternating with beta-agonist inhalers every 2 hours.

19. The nurse understands that brain aneurysms are the result of:

A. an autoimmune process.
B. damage to the seventh cranial nerve (facial).
C. dilation of a cerebral artery.
D. otitis media or tooth abscess.

20. A diagnosis of a brain aneurysm was based on which of the following symptoms demonstrated by Mr. Tyson?

    A. A headache, nuchal rigidity, and a stiff back and legs

    B. Insidious mental deterioration

    C. Severe hypertension, bradycardia, and a pounding headache

    D. Unilateral weakness or paralysis

SITUATION: *Eric Bulson is a 16-year-old high-school student admitted to the pediatric unit with a diagnosis of epilepsy.*

Questions 21 to 24 refer to this situation.

21. The nurse anticipates immediate implementation of which intervention if Eric experiences a seizure?

    A. Administration of phenytoin (Dilantin) and diazepam (Valium) by mouth

    B. Emergency intubation and mechanical ventilation

    C. Ensuring side-lying positioning with padded side rails

    D. Placement of a bite block and physical restraints

22. Eric is ordered to receive 300 mg of phenytoin (Dilantin) by mouth every evening. The nurse's teaching plan for Eric includes meticulous oral hygiene because phenytoin:

    A. may cause bleeding of the gums.

    B. may cause hypertrophy of the gums.

    C. may increase bacterial growth.

    D. may increase plaque, which increases cavities.

23. Eric tells the nurse that he's experiencing a generalized tingling sensation and is "smelling roses." The nurse understands that Eric is probably experiencing:

    A. acute alcohol withdrawal.

    B. an acute cerebrovascular accident (CVA).

    C. an aura.

    D. an olfactory hallucination.

24. The nurse sees that Eric has glassy eyes, is staring, and is performing a chewing motion with his mouth. The nurse correctly assumes that he's experiencing a:

    A. generalized tonic-clonic seizure.
    B. partial complex seizure.
    C. simple focal seizure.
    D. temporal lobe seizure.

SITUATION: *Tracey Dell is a 17-year-old female who's admitted to the intensive care unit (ICU) with an acute spinal cord injury.*

Questions 25 to 28 refer to this situation.

25. The nurse understands that the most common causes of spinal cord injury in patients of this age are:

    A. illnesses such as ankylosing spondylitis and rheumatoid arthritis.
    B. motor vehicle accidents, falls, and diving accidents.
    C. penetrating injuries such as gunshot wounds.
    D. sports injuries.

26. During the first 24 to 48 hours after the injury, the nurse anticipates orders from the doctor to administer:

    A. antihypertensive medications and diuretics.
    B. I.V. fluids and blood products.
    C. I.V. fluids and corticosteroids.
    D. I.V. mannitol.

27. Tracey has been diagnosed with paraplegia. The nurse recognizes that one early problem of paraplegia is:

    A. diarrhea.
    B. learning to use mechanical aids.
    C. patient education.
    D. relearning how to control one's bladder.

28. Tracey develops spinal shock, which is manifested by which of the following signs and symptoms?

    A. Hypertension and bradycardia
    B. Hypotension and tachycardia
    C. Ventricular arrhythmias and hypotension
    D. Ventricular arrhythmias and tachypnea

SITUATION: *Rob Chapman is a 32-year-old construction worker admitted to the emergency department with complete transection of the spinal cord at the fifth thoracic level (T5).*

Questions 29 to 33 refer to this situation.

29. The nurse anticipates which of the following assessment findings regarding Mr. Chapman's voluntary motor activity and sensation?

    A. Complete loss of voluntary motor activity and sensation below the level of injury

    B. Complete loss of voluntary motor activity below the level of injury, with variable sensory loss below the level of injury

    C. Complete loss of voluntary motor activity on one side of the body, with loss of sensation on the contralateral side below the level of injury

    D. Variable loss of voluntary motor activity and sensation below the level of injury

30. Mr. Chapman complains of a severe headache and is extremely anxious. The nurse checks his vital signs and finds him to have a heart rate of 55 beats/minute and a blood pressure of 250/108 mm Hg. The nurse should also assess for:

    A. fecal incontinence.

    B. presence of a Babinski's reflex.

    C. presence of bowel sounds.

    D. urinary catheter patency.

31. Three days later, Mr. Chapman is informed that he will be a paraplegic. He asks the nurse if he will be able to have sexual intercourse. The most knowledgeable response to this question would be:

    A. "It's too early to make this assumption."

    B. "No, I really don't think you will ever have sexual intercourse again."

    C. "This shouldn't be discussed this early in your recovery."

    D. "You are a paraplegic and have lost all motor control below your waist."

32. Mr. Chapman may develop the complication of urinary calculi. The major factor that contributes to this condition in paraplegic patients is:

    A. hypoparathyroidism.

    B. inadequate renal function.

    C. increased intake of calcium.

    D. increased loss of calcium from the skeletal system.

33. When planning long-term care for Mr. Chapman, the nurse should understand that rehabilitation:

A. should begin as soon as the patient is admitted to the hospital.
B. should begin as soon as the patient is stabilized.
C. should be left up to the patient.
D. should not begin until the patient is transferred to a rehabilitation facility.

SITUATION: *Jeanne Meador is a 72-year-old grandmother with a history of mild obesity, osteoporosis, and cataracts.*

Questions 34 to 39 refer to this situation.

34. The nurse performs an assessment on Mrs. Meador. The nurse should understand that which of the following symptoms is most characteristic of cataracts?

A. Decreased discrimination between green and blue
B. None — most patients are asymptomatic
C. Painful, sudden bilateral vision loss
D. Painless, progressive vision loss in one eye or both eyes

35. Which the following statements is true regarding the visual changes associated with cataracts?

A. Both eyes typically develop cataracts at the same rate.
B. The loss of vision is experienced as a painless, gradual blurring.
C. The patient is suddenly blind.
D. The patient typically experiences a painful, sudden blurring of vision.

36. Treatment of cataracts generally involves:

A. application of lubricant to eyes four time a day.
B. application of miotic eyedrops twice a day.
C. no treatment — cataracts are a normal part of aging.
D. surgical removal of cataracts.

37. Mrs. Meador is scheduled for outpatient surgical correction of her cataracts. Discharge teaching on this patient should be based on the nurse's knowledge that:

A. cataracts are usually removed with the patient receiving general anesthesia.
B. cataract surgery typically involves a 2- to 3-day hospital stay.
C. surgery can restore about 75% of vision loss by removing the cataract.
D. surgery can restore about 95% of vision loss by removing the cataract.

38. The nurse should include which of the following instructions in Mrs. Meador's discharge teaching?

   A. Bed rest for 1 week
   B. Limited alcohol and nicotine consumption
   C. No behavior modifications are necessary
   D. No bending, straining, lifting, or coughing

39. Which of the following choices are complications of cataract surgery?

   A. Intraocular infarction and uveitis
   B. Intraocular infection and uveitis
   C. Vitreous collapse and intraocular infection
   D. Vitreous prolapse and intraocular infarction

SITUATION: *Bill Weber is a 78-year-old retired mechanic who's admitted from home with severe vertigo and tinnitus. He has a history of prostatic hypertrophy and is hard of hearing in his left ear.*

Questions 40 to 42 refer to this situation.

40. The nurse expects which of the following characteristics to be true regarding normal physiologic changes to Mr. Weber's hearing?

   A. Asymmetrical sensorineural hearing loss
   B. Difficulty hearing high-frequency sounds symmetrically
   C. Progressive hearing loss in the middle ear
   D. Less distortion of sounds symmetrically

41. Mr. Weber is displaying symptoms of:

   A. glaucoma.
   B. delirium.
   C. Mendel's disease.
   D. Ménière's disease.

42. Nursing interventions for Mr. Weber should include:

   A. administration of narcotic agents during severe attacks.
   B. encouragement of frequent independent activity.
   C. encouragement of a high-sodium diet and intake of fluids.
   D. frequent rest periods, with nursing assistance with activity.

SITUATION: *Alice Brockman is a 75-year-old active female experiencing bouts of urinary incontinence.*

Questions 43 to 45 refer to this situation.

43. Mrs. Brockman may be at increased risk of urinary incontinence because of:

  A. decreased bladder capacity.
  B. a dilated urethra.
  C. diuretic use.
  D. increased glomerular filtration rate.

44. Mrs. Brockman is diagnosed with stress incontinence. Initial treatment of this bladder condition involves:

  A. abdominal wall exercises.
  B. beta-adrenergic agonists.
  C. pelvic floor exercises (Kegel exercises).
  D. temporary urinary catheter placement.

45. Mrs. Brockman's incontinence places her at risk for which of the following problems?

  A. Decreased mobility
  B. Falls
  C. Fecal incontinence
  D. Social isolation

SITUATION: *Henry Jasper is an 80-year-old man who's admitted with a diagnosis of dementia and malnutrition.*

Questions 46 to 50 refer to this situation.

46. The doctor orders a series of laboratory tests to determine whether Mr. Jasper's dementia is treatable. The nurse understands that the most common cause of dementia in this patient population is:

  A. acquired immunodeficiency syndrome (AIDS).
  B. Alzheimer's disease.
  C. brain tumors.
  D. vascular disease.

47. Mr. Jasper is in the middle stage of dementia. Observable behaviors of this stage include:

  A. difficulty concentrating and learning new material.
  B. easy distraction requiring constant supervision of patient.
  C. getting lost easily in unfamiliar places.
  D. an inability to feed oneself and dependence on nonverbal communication.

48. The most likely treatment for Mr. Jasper will include:

    A. the implementation of specific diets and exercises.

    B. the treatment of symptoms as well as patient and family support.

    C. treatment with lecithin and physostigmine.

    D. the use of physical restraints when the patient is left unsupervised.

49. Nursing measures aimed at correcting Mr. Jasper's nutritional status include:

    A. providing excessive stimulation at mealtimes to stimulate hunger.

    B. providing meals in calm surroundings without distractions.

    C. providing meals with many choices.

    D. strictly enforcing mealtimes.

50. Measures to support Mr. Jasper's family should include:

    A. discouraging planning for the future because outcomes in this patient population are unpredictable.

    B. encouraging planning for the future, including financial planning and nursing home application.

    C. encouraging isolation and minimal patient activities.

    D. encouraging as much patient stimulation and as many activities as possible.

SITUATION: *Clint Wilson is a 55-year-old man with Parkinson's disease.*

Questions 51 to 56 refer to this situation

51. The nurse understands that the etiology of Parkinson's disease is:

    A. a syndrome of nerve cell loss in the basal ganglia and neurotransmitter deficiencies.

    B. demyelination of the white matter of the brain and the spinal cord.

    C. progressive spinal muscular atrophy and progressive bulbar palsy.

    D. progressive weakness and fatigue of the skeletal muscle system.

52. Mr. Wilson may manifest which of the following clinical findings?

    A. Ataxia, diplopia, and impaired speech

    B. Ptosis, diplopia, and eye squint

    C. Resting tremors, bradykinesia, and propulsive gait

    D. Vertigo, nausea, and nystagmus

53. The nurse anticipates the use of which drug therapy combination for Mr. Wilson?

    A. Alpha-adrenergic blockers and anticholinergics
    B. Anticholinergics and corticosteroids
    C Anticholinergics and dopaminergic/dopamine-agonist agents
    D. Beta-adrenergic blockers and anticholinergics

54. Mr. Wilson's signs and symptoms are the result of:

    A. a degenerative disease of the nerve impulses.
    B. a depleted concentration of dopamine.
    C. an excess of acetylcholine.
    D. a lack of depolarization.

55. Physical assessment data for Mr. Wilson should include:

    A. cardiopulmonary assessment, deep tendon reflexes, and mental status.
    B. cardiopulmonary assessment, functional ability, and activities of daily living.
    C. range of motion, deep tendon reflexes, and mental status.
    D. motor system, functional ability, and activities of daily living.

56. The nurse provides education and support to Mr. Wilson's family. Realistic goals for Mr. Wilson include:

    A. beginning recovery from disease with the start of treatment.
    B. facilitating adaptation to activities of daily living and self-care.
    C. preparing for imminent death.
    D. preparing to be bedridden within 2 years.

SITUATION: *Arthur Pruitt is a 78-year-old man who's admitted with left-sided weakness, left visual field deficit, and left neglect. A computed tomography scan reveals a right hemispheric infarct.*

Questions 57 to 61 refer to this situation.

57. The nurse anticipates that the most dramatic improvements in functional status for Mr. Pruitt will occur at what point following his stroke?

    A. During the first week
    B. Over the course of the first year
    C. Over the next 2 to 5 years
    D. Within the first 6 months

58. Interventions specific for Mr. Pruitt's cognitive defects include:

   A. capitalization of learning activities as they arise throughout the day.
   B. discouragement of family participation in teaching sessions to promote patient independence.
   C. enforcement of a consistent routine over short time periods.
   D. verbal reinforcement of patient weaknesses and future goals at the end of each session.

59. Which of the following nursing actions would best help Mr. Pruitt overcome his spatial deficit?

   A. Approach the patient from the affected side frequently.
   B. Encourage the patient to keep his belongings neatly out of his field of vision.
   C. Keep the patient isolated for safety.
   D. Support and promote the patient's undisturbed field of vision.

60. Once Mr. Pruitt understands the presence of his visual field deficit, a useful technique the nurse can teach him is:

   A. memorizing.
   B. minimizing.
   C. scaling.
   D. scanning.

61. A leading cause of mortality in this patient population during the rehabilitative period includes:

   A. chronic obstructive pulmonary disease exacerbation.
   B. development of autoimmune disease.
   C. extension of underlying neurological injury.
   D. myocardial infarction (MI).

SITUATION: *Alyssa McBride is a 35-year-old accountant admitted to the hospital with Bell's palsy.*

   Questions 62 to 64 refer to this situation.

62. The nurse anticipates which of the following findings when performing a physical examination on Mrs. McBride?

   A. Bilateral facial droop
   B. Bilateral lower extremity weakness
   C. Unilateral facial droop with flat nasolabial fold
   D. Unilateral weakness with ipsilateral loss of sensation

63. The nurse is aware that Bell's palsy affects which cranial nerve?

   A. Second cranial nerve (optic)
   B. Third cranial nerve (oculomotor)
   C. Fourth cranial nerve (trochlear)
   D. Seventh cranial nerve (facial)

64. The nurse's interventions for Mrs. McBride should include:

   A. careful preoperative instructions.
   B. meticulous eye care and nutritional supplementation.
   C. splinting of the affected extremities.
   D. the use of a walker to aid in ambulation.

SITUATION: *George Nager is a 67-year-old patient who has been diagnosed with chronic open-angle glaucoma.*

   Questions 65 to 67 refer to this situation.

65. The nurse understands that chronic open-angle glaucoma:

   A. is the most common form of glaucoma.
   B. manifests as a rapid loss of peripheral vision.
   C. manifests as a slow loss of central vision.
   D. occurs most frequently in people of Asian heritage.

66. Mr. Nager asks the nurse, "What causes my glaucoma?" The nurse's best response is based on her understanding that the pathophysiology of chronic open-angle glaucoma includes:

   A. inadequate drainage of aqueous fluid from the eye.
   B. overproduction of aqueous humor.
   C. overproduction of aqueous humor with or without adequate drainage of aqueous fluid from the eye.
   D. a shallow anterior eye chamber.

67. Mr. Nager asks the nurse whether medical or surgical interventions can restore vision. The nurse's response regarding Mr. Nager's vision following medical treatment and surgical trabeculectomy or laser trabeculoplasty is based on her knowledge that:

   A. further vision loss is prevented, but previous vision loss cannot be restored.
   B. peripheral vision loss can be restored, but central vision remains lost.
   C. visual acuity continues to deteriorate slowly.
   D. visual acuity is restored.

SITUATION: *Edith Kelly is a 75-year-old woman with a history of hypertension and cervical cancer. She's also hard of hearing in her left ear.*

Questions 68 to 72 refer to this situation.

68. To best facilitate communication with Mrs. Kelly, the nurse should:

    A. begin conversation with the most pressing topics.
    B. quickly shift conversation from topic to topic.
    C. reduce distracting background noise.
    D. repeat misunderstood sentences as needed.

69. Mrs. Kelly most likely has which type of hearing loss?

    A. Conductive
    B. Fluctuating
    C. Neural
    D. Presbycusis

70. Mrs. Kelly receives a hearing aid. What instructions should the nurse give to Mrs. Kelly's family?

    A. Check hearing aid function after placing the aid in the patient's ear.
    B. Remove the aid frequently during the day.
    C. Replace the battery if no squealing occurs with the volume at its highest level.
    D. Speak away from the patient to see if she hears you.

71. The nurse is aware that which clue indicates that Mrs. Kelly's hearing aid isn't working?

    A. The patient is able to distinguish voices despite a noisy environment.
    B. The patient is unable to hear high-pitched voices.
    C. The patient is observed engaging in group conversations.
    D. The patient is observed reminiscing.

72. The nurse teaches Mrs. Kelly that the functional parts of her hearing aid include:

    A. an electric microphone that transmits sounds to a receiver.
    B. a microphone that converts sounds to electrical signals, which are stored for further use.
    C. a microphone that converts sounds to electrical signals, which are transmitted to a receiver.
    D. a microphone that transmits sounds to an electrical receiver.

SITUATION: *Lucy Crittendon is a 55-year-old woman who's being admitted to the medical floor unit with trigeminal neuralgia.*

Questions 73 to 76 refer to this situation.

73. The nurse anticipates that Mrs. Crittendon will demonstrate which of the following major complaints?

    A. Excruciating, intermittent, paroxysmal facial pain
    B. Mildly painful unilateral eye twitching
    C. Unilateral facial droop
    D. Unilateral painless eye spasm

74. The nurse is aware that Mrs. Crittendon's trigeminal neuralgia is affecting which of her cranial nerves?

    A. Second cranial nerve
    B. Fifth cranial nerve
    C. Fourth cranial nerve
    D. Seventh cranial nerve

75. The nurse anticipates that Mrs. Crittendon will require which of the following interventions?

    A. High-dose corticosteroids followed by anticonvulsant drug therapy
    B. High-dose corticosteroids followed by surgical intervention
    C. Immediate anticonvulsant drug therapy followed by surgical intervention
    D. Immediate surgical intervention followed by anticonvulsant drug therapy

76. Mrs. Crittendon requires surgical intervention for her trigeminal neuralgia. The nurse anticipates providing Mrs. Crittendon with which of the following postoperative instructions?

    A. Application of an eye patch for at least a week
    B. Application of lubricating eyedrops as needed
    C. Avoidance of food or drink for 24 hours
    D. Avoidance of oral care for 24 hours

ANSWER SHEET

A B C D		A B C D		A B C D
1 ○ ○ ○ ○		27 ○ ○ ○ ○		53 ○ ○ ○ ○
2 ○ ○ ○ ○		28 ○ ○ ○ ○		54 ○ ○ ○ ○
3 ○ ○ ○ ○		29 ○ ○ ○ ○		55 ○ ○ ○ ○
4 ○ ○ ○ ○		30 ○ ○ ○ ○		56 ○ ○ ○ ○
5 ○ ○ ○ ○		31 ○ ○ ○ ○		57 ○ ○ ○ ○
6 ○ ○ ○ ○		32 ○ ○ ○ ○		58 ○ ○ ○ ○
7 ○ ○ ○ ○		33 ○ ○ ○ ○		59 ○ ○ ○ ○
8 ○ ○ ○ ○		34 ○ ○ ○ ○		60 ○ ○ ○ ○
9 ○ ○ ○ ○		35 ○ ○ ○ ○		61 ○ ○ ○ ○
10 ○ ○ ○ ○		36 ○ ○ ○ ○		62 ○ ○ ○ ○
11 ○ ○ ○ ○		37 ○ ○ ○ ○		63 ○ ○ ○ ○
12 ○ ○ ○ ○		38 ○ ○ ○ ○		64 ○ ○ ○ ○
13 ○ ○ ○ ○		39 ○ ○ ○ ○		65 ○ ○ ○ ○
14 ○ ○ ○ ○		40 ○ ○ ○ ○		66 ○ ○ ○ ○
15 ○ ○ ○ ○		41 ○ ○ ○ ○		67 ○ ○ ○ ○
16 ○ ○ ○ ○		42 ○ ○ ○ ○		68 ○ ○ ○ ○
17 ○ ○ ○ ○		43 ○ ○ ○ ○		69 ○ ○ ○ ○
18 ○ ○ ○ ○		44 ○ ○ ○ ○		70 ○ ○ ○ ○
19 ○ ○ ○ ○		45 ○ ○ ○ ○		71 ○ ○ ○ ○
20 ○ ○ ○ ○		46 ○ ○ ○ ○		72 ○ ○ ○ ○
21 ○ ○ ○ ○		47 ○ ○ ○ ○		73 ○ ○ ○ ○
22 ○ ○ ○ ○		48 ○ ○ ○ ○		74 ○ ○ ○ ○
23 ○ ○ ○ ○		49 ○ ○ ○ ○		75 ○ ○ ○ ○
24 ○ ○ ○ ○		50 ○ ○ ○ ○		76 ○ ○ ○ ○
25 ○ ○ ○ ○		51 ○ ○ ○ ○		
26 ○ ○ ○ ○		52 ○ ○ ○ ○		

ANSWERS AND RATIONALES

1. *Correct answer:* **B**

Damage to the third cranial nerve (oculomotor) may cause dilation of the pupil. The second cranial nerve (optic) is responsible for acuity and visual field, the fourth cranial nerve (trochlear) is responsible for extraocular movements, and the seventh cranial nerve (facial) innervates the eyebrow and upper eyelid.

> *Nursing process step:* Assessment
> *Client needs category:* Physiological integrity
> *Client needs subcategory:* Physiological adaptation
> *Taxonomic level:* Knowledge

2. *Correct answer:* **C**

For the body to absorb calcium, pressure must be placed on the long bones. The tilt table places the patient in a standing position, which puts pressure on the long bones. The tilt table is raised slowly to prevent hypotension, but preventing hypotension isn't the purpose of the table. The table also increases circulation to the extremities as well as helps prevent pressure sores; these benefits also aren't the main purpose of the table.

> *Nursing process step:* Implementation
> *Client needs category:* Physiological integrity
> *Client needs subcategory:* Reduction of risk potential
> *Taxonomic level:* Comprehension

3. *Correct answer:* **B**

Huntington's disease is a hereditary disease in which degeneration of the basal ganglia and cerebral cortex causes chronic progressive chorea (muscle twitching) and mental deterioration, ending in dementia. Huntington's disease usually strikes people ages 25 to 55. The cause is unknown, the onset is insidious, and death usually occurs 10 to 15 years after the onset. Myasthenia gravis produces progressive muscle weakness, MS causes progressive demyelination of the white matter of the spinal cord and the brain, and Guillain-Barré syndrome exhibits a progressive polyneuritis, causing ascending muscular weakness and paralysis.

> *Nursing process step:* Assessment
> *Client needs category:* Physiological integrity
> *Client needs subcategory:* Physiological adaptation
> *Taxonomic level:* Comprehension

4. *Correct answer:* **A**

Seizures may occur at any time following a head injury, but the onset of seizures specifically tends to occur 6 months to 2 years posttrauma.

> *Nursing process step:* Implementation
> *Client needs category:* Physiological integrity
> *Client needs subcategory:* Physiological adaptation
> *Taxonomic level:* Knowledge

5. *Correct answer:* **B**

Quadriplegia results from an injury to the spinal cord anywhere between levels C1 to C8. Injuries to the spinal cord at levels below C8 can cause lower level paralysis such as paraplegia.

> *Nursing process step:* Assessment
> *Client needs category:* Physiological integrity
> *Client needs subcategory:* Physiological adaptation
> *Taxonomic level:* Knowledge

6. *Correct answer:* **C**

The upper motor neuron originates in the cerebral cortex and terminates in the anterior horn cell in the spinal cord. Reflexes return following an injury, resulting in spastic paralysis. Flaccid paralysis occurs following lower motor neuron damage because the lower motor neuron begins at the anterior horn cell and connects to the motor side of the reflex arc.

> *Nursing process step:* Assessment
> *Client needs category:* Physiological integrity
> *Client needs subcategory:* Physiological adaptation
> *Taxonomic level:* Comprehension

7. *Correct answer:* **A**

The elasticity of the eye's lens and the pupillary light reflex decline with age. Decreased tearing of the eyes also is normal in the elderly. Strabismus is an abnormal finding.

> *Nursing process step:* Assessment
> *Client needs category:* Health promotion and maintenance
> *Client needs subcategory:* Growth and development through the life span
> *Taxonomic level:* Knowledge

8. *Correct answer:* **A**

Physical changes associated with aging include decreased physical strength and endurance, decreased muscular coordination, decreased tolerance to cold, and a tendency to lose weight.

> *Nursing process step:* Assessment
> *Client needs category:* Health promotion and maintenance
> *Client needs subcategory:* Growth and development through the life span
> *Taxonomic level:* Knowledge

9. *Correct answer:* **B**

Psychogenic pain is related to emotional factors and has no anatomic or physiological explanation. Pain due to an amputation is called phantom pain, pain originating in the internal organs is visceral pain, and pain originating in the central nervous system is central pain.

> *Nursing process step:* Assessment
> *Client needs category:* Physiological integrity
> *Client needs subcategory:* Coping and adaptation
> *Taxonomic level:* Knowledge

10. *Correct answer:* **C**

Major fears of elderly people include fear of physical and economic dependency as well as fear of chronic illness and fear of loneliness.

> *Nursing process step:* Assessment
> *Client needs category:* Physiological integrity
> *Client needs subcategory:* Coping and adaptation
> *Taxonomic level:* Comprehension

11. *Correct answer:* **C**

When a patient has had a cerebrovascular accident (CVA) in the left hemisphere of the brain, the paresis will be on the right side of the body. Because the pyramidal motor tracts cross over at the medulla oblongata, the right side of the body and the right side of the face will have paresis with left-hemisphere damage. The four extremities wouldn't be affected by a left-hemisphere CVA, and neither would the left side of the body.

> *Nursing process step:* Assessment
> *Client needs category:* Physiological integrity
> *Client needs subcategory:* Physiological adaptation
> *Taxonomic level:* Comprehension

12. *Correct answer:* **D**

Aphasia describes the absence or impairment of speech when the patient can still communicate through gestures. Acephalia describes the congenital absence of the head. An accoucheur is an obstetrician or midwife. Akathisia is the inability to sit because the thought of doing so causes severe anxiety.

> *Nursing process step:* Assessment
> *Client needs category:* Physiological integrity
> *Client needs subcategory:* Physiological adaptation
> *Taxonomic level:* Knowledge

13. *Correct answer:* **C**

Long-term planning for a patient with aphasia would include consulting with the patient's doctor concerning a speech therapy program. In Mr. Simpson's case, it's too soon to know whether his disability will be permanent. Expecting patients with aphasia to respond verbally is emotionally upsetting for the patients, For patients with complete motor aphasia, associating words with objects serves no purpose.

> *Nursing process step:* Planning
> *Client needs category:* Physiological integrity
> *Client needs subcategory:* Physiological adaptation
> *Taxonomic level:* Application

14. *Correct answer:* **D**

Using gestures to accompany the spoken word will help the patient better understand the message being communicated. Conversing slowly and giving directions one at a time will allow the patient time to understand what is being said. Speaking in a louder-than-usual tone of voice won't help a stroke patient; the patient has a speech problem, not a hearing problem.

> *Nursing process step:* Implementation
> *Client needs category:* Physiological integrity
> *Client needs subcategory:* Physiological adaptation
> *Taxonomic level:* Application

15. *Correct answer:* **A**

Guillain-Barré syndrome is an acute, rapidly progressive, and potentially fatal form of polyneuritis that causes muscle weakness and mild distal sensory loss. In contrast, Huntington's disease is a hereditary disease in which degeneration of the cerebral cortex and basal ganglia causes chronic progressive chorea (muscle twitching) and mental deterioration, ending in dementia. MS causes progressive demyelination of the white matter of the brain and the spinal cord, and myasthenia gravis produces progressive weakness of the skeletal muscles — including respiratory muscles — which become weaker with exercise.

Nursing process step: Assessment
Client needs category: Physiological integrity
Client needs subcategory: Physiological adaptation
Taxonomic level: Comprehension

16. *Correct answer:* **C**

Patients with Guillain-Barré syndrome may suffer from progressive paralysis that includes the respiratory muscles and may require intubation and mechanical ventilation during the acute phase of their disease. Pyridostigmine (Mestinon) is an anticholinesterase drug used to treat myasthenia gravis; high-dose steroids and wide-spectrum antibiotics aren't indicated for the treatment of Guillain-Barré syndrome.

Nursing process step: Planning
Client needs category: Physiological integrity
Client needs subcategory: Physiological adaptation
Taxonomic level: Application

17. *Correct answer:* **C**

Treatment of Guillain-Barré syndrome is supportive and includes intubation with mechanical ventilation, nutritional supplementation, and rigorous attention to hygiene and skin care.

Nursing process step: Implementation
Client needs category: Physiological integrity
Client needs subcategory: Physiological adaptation
Taxonomic level: Knowledge

18. *Correct answer:* **B**

Hypotensive agents, steroids, and sedatives are utilized to prevent rupture of the aneurysm or to reduce bleeding if a cerebrovascular accident has occurred. Patients are typically kept on bed rest with a quiet room and no stimulants. Phenytoin and phenobarbitol are used to treat seizures. Terbutaline and beta-agonists would increase blood pressure and are contraindicated in this patient.

Nursing process step: Planning
Client needs category: Physiological integrity
Client needs subcategory: Pharmacological and parenteral therapies
Taxonomic level: Application

19. *Correct answer:* C

A brain aneurysm is a dilation of a cerebral artery, which is caused secondary to weakness in the vessel wall. The vessel may rupture and cause a subarachnoid bleed. Autoimmune diseases that cause neurological damage include multiple sclerosis, myasthenia gravis, and Lou Gehrig's disease. Otitis media, in rare instances, may cause a brain abscess, and damage to the seventh cranial nerve (facial) results in Bell's palsy.

> *Nursing process step:* Assessment
> *Client needs category:* Physiological integrity
> *Client needs subcategory:* Physiological adaptation
> *Taxonomic level:* Knowledge

20. *Correct answer:* **A**

Signs of a brain aneurysm may include a headache, nuchal rigidity, and a stiff back and legs, progressing to altered consciousness and coma in severe cases. Unilateral weakness or paralysis is most characteristic of a hemispheric cerebrovascular accident. Insidious mental deterioration indicates Alzheimer's disease; the triad of severe hypertension, bradycardia, and a headache may signal autonomic hyperreflexia, which is an acute complication of spinal cord injury.

> *Nursing process step:* Assessment
> *Client needs category:* Physiological integrity
> *Client needs subcategory:* Physiological adaptation
> *Taxonomic level:* Comprehension

21. *Correct answer:* C

During a seizure, the patient should be positioned on his side to prevent aspiration if he vomits. Padding the side rails prevents injury during generalized tonic-clonic seizures. With correct positioning, the patient should maintain a patent airway; hence, intubation should not be necessary. Items — including bite blocks and medications — should never be forced into the patient's mouth during a seizure.

> *Nursing process step:* Planning
> *Client needs category:* Physiological integrity
> *Client needs subcategory:* Reduction of risk potential
> *Taxonomic level:* Application

22. *Correct answer:* **B**

Because phenytoin (Dilantin) may cause hypertrophy of the gums, meticulous oral hygiene is required to reduce inflammation and prevent infection of the gums. Phenytoin doesn't cause an increase in bacterial growth or plaque formation, nor does it cause bleeding of the gums.

Nursing process step: Implementation
Client needs category: Physiological integrity
Client needs subcategory: Pharmacological and parenteral therapies
Taxonomic level: Comprehension

23. *Correct answer:* **C**

An aura frequently precedes an epileptic seizure and may manifest as vague psychic discomfort or specific aromas. Patients experiencing auras aren't having a CVA, experiencing substance withdrawal, or hallucinating.

Nursing process step: Assessment
Client needs category: Physiological integrity
Client needs subcategory: Physiological adaptation
Taxonomic level: Comprehension

24. *Correct answer:* **C**

A simple focal seizure begins in one part of the cerebral cortex and remains localized, resulting in specific isolated behaviors and brief lapses of attention. Generalized seizures start in one part of the cortex but spread throughout the brain, resulting in widespread abnormal muscle activity that's usually in the form of generalized tonic-clonic seizures. Partial complex seizures start in one part of the cortex but generalize throughout part or all of the brain, resulting in abnormal motor activity. Temporal lobe seizures manifest as abnormal behavioral outbursts and are treated as psychiatric disorders.

Nursing process step: Assessment
Client needs category: Physiological integrity
Client needs subcategory: Physiological adaptation
Taxonomic level: Comprehension

25. *Correct answer:* **B**

The most common causes of spinal cord injury in teenagers and young adults are motor vehicle accidents, falls, and diving accidents. Arthritic conditions may occasionally cause spinal cord injury secondary to narrowing of the spinal canal in the elderly population. Penetrating injuries such as gunshot wounds occur less frequently (and mostly in young men). Sports injuries also are less likely to cause spinal cord injury.

Nursing process step: Assessment
Client needs category: Physiological integrity
Client needs subcategory: Physiological adaptation
Taxonomic level: Knowledge

26. *Correct answer:* **C**

In the early stages of spinal cord injury, patients experience a relative hypovolemia, with inflammation contributing to secondary injury to the damaged spinal cord region. Patients are treated with I.V. fluids and corticosteroids. Antihypertensive medications and diuretics are typically contraindicated in these patients. Blood products are required only if the patient has experienced blood loss, and I.V. mannitol is indicated for cerebral edema, not for a spinal cord injury.

> *Nursing process step:* Planning
> *Client needs category:* Physiological integrity
> *Client needs subcategory:* Pharmacological and parenteral therapies
> *Taxonomic level:* Application

27. *Correct answer:* **B**

Learning to use mechanical aids is an early problem for a paraplegic. Most paraplegics are young adults and are accustomed to leading active, independent lives. They must relearn many motor skills by using mechanical aids. Education isn't a problem for these patients unless they've suffered concomitant head injury. Paraplegics have no bladder control, so they need to learn how to catheterize themselves. Constipation, rather than diarrhea, is a major problem because of the paralytic ileus.

> *Nursing process step:* Assessment
> *Client needs category:* Physiological integrity
> *Client needs subcategory:* Physiological adaptation
> *Taxonomic level:* Comprehension

28. *Correct answer:* **B**

Spinal shock occurs in patients with complete transection of the spinal cord. Impairment of the vasomotor mechanism (nerves having muscular control of the blood vessel walls) causes a drop in blood pressure and an increased heart rate. The tachycardia is a compensatory mechanism. The other answers are incorrect.

> *Nursing process step:* Assessment
> *Client needs category:* Physiological integrity
> *Client needs subcategory:* Physiological adaptation
> *Taxonomic level:* Comprehension

29. *Correct answer:* **A**

Following a complete spinal cord transection, all voluntary motor activity and sensation below the level of injury is lost. Variable loss of voluntary motor activity and sensation occurs following partial transection of the spinal cord. Loss of voluntary motor activity on one side of the body with loss of sensation on the other side below the level of injury is characteristic of Brown-Séquard syndrome.

Nursing process step: Assessment
Client needs category: Physiological integrity
Client needs subcategory: Physiological adaptation
Taxonomic level: Comprehension

30. *Correct answer:* **D**

The patient is complaining of symptoms of autonomic dysreflexia, which consists of the triad of hypertension, bradycardia, and a headache. Major causes of autonomic dysreflexia include urinary bladder distention and fecal impaction. Checking the patency of the urinary catheter will check for bladder distention. Fecal incontinence isn't related to this syndrome, and the presence or absence of bowel sounds and lower extremity reflexes doesn't help in the diagnosis or management of this condition.

Nursing process step: Assessment
Client needs category: Physiological integrity
Client needs subcategory: Physiological adaptation
Taxonomic level: Application

31. *Correct answer:* **A**

The short span of 3 days after a spinal cord injury is too early to determine a male patient's ability to have sexual intercourse.

Nursing process step: Implementation
Client needs category: Physiological integrity
Client needs subcategory: Coping and adaptation
Taxonomic level: Application

32. *Correct answer:* **D**

Paraplegic patients can't put pressure—which is necessary for calcium absorption—on their long bones. This causes the skeletal system to lose calcium, a situation that causes an increase in serum calcium levels and can cause renal calculi. Renal calculi are usually not caused by hypoparathyroidism, inadequate renal function, or an increased intake of calcium.

Nursing process step: Assessment
Client needs category: Physiological integrity
Client needs subcategory: Reduction of risk potential
Taxonomic level: Comprehension

33. *Correct answer:* **B**

Rehabilitation should begin as soon as the patient is stabilized after a spinal cord injury. It can't be left up to the patient, particularly because he may be depressed and uninterested in rehabilitation early in recovery. Valuable time may be lost if

rehabilitation isn't started until the patient reaches a specialized facility. If rehabilitation is started during the acute phase of injury, activities may seriously in-
jure the patient if his injury isn't stable.

> *Nursing process step:* Planning
> *Client needs category:* Physiological integrity
> *Client needs subcategory:* Physiological adaptation
> *Taxonomic level:* Comprehension

34. *Correct answer:* **D**

Cataracts are caused by clouding or opacity of the lens that leads to eventual loss
of sight. Most patients have painless progressive vision loss, sometimes with increased glare from bright lights. Cataracts may occur in one eye or both eyes.

> *Nursing process step:* Assessment
> *Client needs category:* Physiological integrity
> *Client needs subcategory:* Physiological adaptation
> *Taxonomic level:* Comprehension

35. *Correct answer:* **B**

Typically, a patient with cataracts experiences a painless, gradual loss of vision.
Although both eyes may develop cataracts, the cataracts usually develop at different rates.

> *Nursing process step:* Assessment
> *Client needs category:* Physiological integrity
> *Client needs subcategory:* Physiological adaptation
> *Taxonomic level:* Knowledge

36. *Correct answer:* **D**

Surgical removal of cataracts is performed when vision loss interferes with the
patient's functional status. Lubricants and miotic eyedrops don't improve
cataracts. Cataracts occur in many older adults but aren't part of normal aging.

> *Nursing process step:* Implementation
> *Client needs category:* Physiological integrity
> *Client needs subcategory:* Physiological adaptation
> *Taxonomic level:* Comprehension

37. *Correct answer:* **D**

Surgery can restore about 95% of vision loss by removing the cataract. The procedure is generally performed under local anesthesia in an ambulatory surgical
center, with patients returning home 2 to 3 hours postoperatively.

Nursing process step: Planning
Client needs category: Physiological integrity
Client needs subcategory: Physiological adaptation
Taxonomic level: Knowledge

38. *Correct answer:* **D**

Patients must be careful to not increase intraocular pressure by bending, straining, lifting, or coughing. Bed rest isn't necessary. Alcohol and nicotine don't directly affect recovery.

Nursing process step: Implementation
Client needs category: Physiological integrity
Client needs subcategory: Reduction of risk potential
Taxonomic level: Application

39. *Correct answer:* **B**

Intraocular infection and uveitis — together with vitreous prolapse and hyphema — are major complications of cataract surgery.

Nursing process step: Assessment
Client needs category: Physiological integrity
Client needs subcategory: Physiological adaptation
Taxonomic level: Comprehension

40. *Correct answer:* **B**

Due to presbycusis (sensorineural hearing loss), the ability to hear high-frequency sounds is lost first. Progressive hearing loss occurs in the inner ear and causes sounds to become more distorted, particularly in crowded or noisy surroundings. Asymmetrical sensorineural hearing loss is abnormal.

Nursing process step: Assessment
Client needs category: Health promotion and maintenance
Client needs subcategory: Growth and development through the life span
Taxonomic level: Knowledge

41. *Correct answer:* **D**

Ménière's disease is a labyrinthine dysfunction. Manifestations include severe vertigo, tinnitus, and sensorineural hearing loss. Delirium is sometimes called confusional state. It usually begins with confusion and progresses to disorientation and changes in the level of consciousness. Glaucoma refers to a group of diseases that are generally characterized by visual field loss caused by damage to the optic nerve. Mendel's disease isn't a disease.

Nursing process step: Assessment
Client needs category: Physiological integrity
Client needs subcategory: Physiological adaptation
Taxonomic level: Comprehension

42. *Correct answer:* **D**

During severe attacks of Ménière's disease, patients should rest in a position that minimizes their vertigo. Patients also should request assistance with activities to prevent injury from falls. Treatment includes administration of mild nonnarcotic sedatives, diuretics, and a sodium-restricted diet.

Nursing process step: Implementation
Client needs category: Physiological integrity
Client needs subcategory: Reduction of risk potential
Taxonomic level: Comprehension

43. *Correct answer:* **A**

Bladder capacity decreases to 200 ml or less in postmenopausal women, causing urinary frequency. The glomerular filtration rate decreases in elderly people. Based on stated history, Mrs. Brockman doesn't have a dilated urethra and she doesn't take a diuretic.

Nursing process step: Assessment
Client needs category: Health promotion and maintenance
Client needs subcategory: Growth and development through the life span
Taxonomic level: Comprehension

44. *Correct answer:* **C**

Pelvic floor exercises (Kegel exercises) are the first appropriate intervention for stress incontinence. These exercises are used to strengthen perianal and sphincter muscle control. Alpha-adrenergic agonists are occasionally used as adjuncts to pelvic floor exercises. Placement of a urinary catheter isn't indicated and increases the risk of infection in these patients. Abdominal wall exercises are of no use in the treatment of stress incontinence.

Nursing process step: Implementation
Client needs category: Physiological integrity
Client needs subcategory: Physiological adaptation
Taxonomic level: Comprehension

45. *Correct answer:* **D**

Because of the frequent need to urinate, the patient may limit social activities and suffer from social isolation and lowered self-image.

Nursing process step: Evaluation
Client needs category: Physiological integrity
Client needs subcategory: Coping and adaptation
Taxonomic level: Analysis

46. *Correct answer:* B

Alzheimer's disease is the most common cause of dementia in the elderly population. AIDS, brain tumors, and vascular disease are all less common causes of progressive loss of mental function in elderly patients.

Nursing process step: Assessment
Client needs category: Physiological integrity
Client needs subcategory: Physiological adaptation
Taxonomic level: Knowledge

47. *Correct answer:* B

During the middle stage of dementia, patients require constant supervision to
ensure safety and are easily distracted during activities. Difficulties with concentration and learning are common in early dementia, whereas patients in the late
stages of dementia become nonverbal and require complete care.

Nursing process step: Assessment
Client needs category: Physiological integrity
Client needs subcategory: Physiological adaptation
Taxonomic level: Comprehension

48. *Correct answer:* B

Care of the patient with Alzheimer's disease is symptomatic and supportive, including patient and family counseling and education. Treatment with lecithin
and physostigmine hasn't proven useful, and the use of physical restraints is ethi-
cally unacceptable.

Nursing process step: Planning
Client needs category: Physiological integrity
Client needs subcategory: Physiological adaptation
Taxonomic level: Application

49. *Correct answer:* B

Meals provided in a quiet environment with minimal distractions are most helpful. Offering too many food choices, providing excessive simulation at mealtimes,
and strictly enforcing mealtimes can lead to anger and lack of cooperation in this
patient population.

Nursing process step: Implementation
Client needs category: Physiological integrity
Client needs subcategory: Basic care and comfort
Taxonomic level: Application

50. *Correct answer:* **B**

Families of patients with Alzheimer's disease should be encouraged to plan ahead for progressive patient decline in functioning. This includes financial planning and nursing home or hospice applications. Both overstimulation and understimulation aren't helpful to patients with Alzheimer's disease.

Nursing process step: Implementation
Client needs category: Physiological integrity
Client needs subcategory: Coping and adaptation
Taxonomic level: Knowledge

51. *Correct answer:* **A**

A syndrome of nerve cell loss in the basal ganglia and neurotransmitter deficiencies causes the specific motor problems noted in Parkinson's disease. Demyelination of the white matter of the brain and the spinal cord results in multiple sclerosis; progressive spinal muscular atrophy and progressive bulbar palsy are hallmarks of amyotrophic lateral sclerosis; and progressive weakness and fatigue of the skeletal muscle system is caused by myasthenia gravis.

Nursing process step: Assessment
Client needs category: Physiological integrity
Client needs subcategory: Physiological adaptation
Taxonomic level: Knowledge

52. *Correct answer:* **C**

Resting tremors, bradykinesia, gait disturbances, and rigidity are characteristics of Parkinson's disease. Ataxia, diplopia, and impaired speech are symptoms of multiple sclerosis; ptosis, diplopia, and eye squint are symptoms of myasthenia gravis; and vertigo, nausea, and nystagmus occur in Meniere's disease.

Nursing process step: Assessment
Client needs category: Physiological integrity
Client needs subcategory: Physiological adaptation
Taxonomic level: Knowledge

53. *Correct answer:* **C**

The combination of anticholinergic agents and dopaminergic/dopamine-agonist agents usually provides the most symptomatic relief for patients with Parkinson's disease.

Nursing process step: Planning
Client needs category: Physiological integrity
Client needs subcategory: Pharmacological and parenteral therapies
Taxonomic level: Application

54. *Correct answer:* **B**

Signs and symptoms of Parkinson's disease are caused by a depleted concentration of dopamine. Parkinson's disease *is* a degenerative disease—it doesn't result from one. Smaller amounts of dopamine result in more excitative effects of acetylcholine but there is no acetylcholine excess. Depolarization isn't a problem.

Nursing process step: Assessment
Client needs category: Physiological integrity
Client needs subcategory: Physiological adaptation
Taxonomic level: Comprehension

55. *Correct answer:* **D**

Detailed assessment of the patient's motor system, functional ability, and ability to perform the activities of daily living serves as the basis for determining nursing interventions. Cardiopulmonary, range-of-motion, deep tendon reflexes, and mental status assessments—although helpful—aren't as specific for Parkinson's disease.

Nursing process step: Assessment
Client needs category: Physiological integrity
Client needs subcategory: Physiological adaptation
Taxonomic level: Comprehension

56. *Correct answer:* **B**

Patients and their families should be provided with support to enable the patient to adapt to activities of daily living and self-care by using special aids and devices. The progression of the disease cannot be stopped, nor is it predictable, so education should focus on fostering independence.

Nursing process step: Planning
Client needs category: Physiological integrity
Client needs subcategory: Coping and adaptation
Taxonomic level: Application

57. *Correct answer:* **D**

The most significant improvements usually occur within 6 months.

Nursing process step: Planning
Client needs category: Physiological integrity
Client needs subcategory: Physiological adaptation
Taxonomic level: Knowledge

58. *Correct answer:* **C**

Learning situations for patients rehabilitating from an acute neurological event should be held in quiet areas, over short time periods, and with enforcement of consistent routines. Inconsistent teaching as opportunities arise through the day is more likely to result in patient frustration and fatigue. Family members should be included in teaching sessions so they can follow similar approaches and consistent routines. Teaching sessions should end on a positive note with reinforcement and highlighting of accomplishments.

> *Nursing process step:* Implementation
> *Client needs category:* Physiological integrity
> *Client needs subcategory:* Physiological adaptation
> *Taxonomic level:* Application

59. *Correct answer:* **D**

Patients with spatial difficulties have difficulty finding their position in space. Supporting and promoting their undisturbed field of vision is helpful. They're unable to locate things on top of or behind other objects, and their confusion increases with social isolation or being approached from their affected side.

> *Nursing process step:* Implementation
> *Client needs category:* Physiological integrity
> *Client needs subcategory:* Physiological adaptation
> *Taxonomic level:* Application

60. *Correct answer:* **D**

When the patient begins to understand the presence of his visual field deficit, the nurse may teach him to compensate with scanning the environment by turning his head and moving his eyes in the direction of his affected visual field. Memorizing, minimizing, or scaling won't help him deal with his visual field defect.

> *Nursing process step:* Implementation
> *Client needs category:* Physiological integrity
> *Client needs subcategory:* Physiological adaptation
> *Taxonomic level:* Knowledge

61. *Correct answer:* **D**

MI is the leading cause of late mortality among patients with completed stroke. Each patient's endurance and medical status must be taken into consideration during the rehabilitation period. Extension of the underlying neurological damage is most likely to occur during the immediate acute phase of cerebral infarction. The remaining answers are incorrect in this context.

Nursing process step: Assessment
Client needs category: Physiological integrity
Client needs subcategory: Reduction of risk potential
Taxonomic level: Knowledge

62. *Correct answer:* **C**

Bell's palsy is a unilateral facial weakness of sudden onset, secondary to paralysis of a cranial nerve. The other symptoms don't occur in Bell's palsy.

Nursing process step: Assessment
Client needs category: Physiological integrity
Client needs subcategory: Physiological adaptation
Taxonomic level: Comprehension

63. *Correct answer:* **D**

Bell's palsy is paralysis of the motor component of the seventh cranial nerve (facial), resulting in facial sag, an inability to close the eyelid or the mouth, drooling, a flat nasolabial fold, and the loss of taste on the affected side of the face.

Nursing process step: Assessment
Client needs category: Physiological integrity
Client needs subcategory: Physiological adaptation
Taxonomic level: Knowledge

64. *Correct answer:* **B**

Protection of the eye is extremely important when the eyelid doesn't function normally. Lubricating drops during the day and ointment at night as well as nutritional supplementation are helpful. Bell's palsy affects only the face, so functional aids aren't necessary. There's no surgical procedure that helps this condition.

Nursing process step: Implementation
Client needs category: Physiological integrity
Client needs subcategory: Reduction of risk potential
Taxonomic level: Application

65. *Correct answer:* **A**

Chronic open-angle glaucoma is the most common form of glaucoma. Risk factors include African-American heritage, hypertension, and diabetes. The condition typically manifests as a slow loss of peripheral vision followed by the eventual loss of central vision.

Nursing process step: Assessment
Client needs category: Physiological integrity
Client needs subcategory: Physiological adaptation
Taxonomic level: Knowledge

66. *Correct answer:* **C**

Chronic open-angle glaucoma is due to an overproduction of aqueous humor with or without drainage of aqueous fluid from the eye. The result of this overproduction is increased intraocular pressure, which causes cupping of the optic disk. Eventual destruction of the nerve fibers in the retina occurs, leading to visual loss in affected areas. A shallow anterior eye chamber may contribute to closed-angle glaucoma.

Nursing process step: Implementation
Client needs category: Physiological integrity
Client needs subcategory: Physiological adaptation
Taxonomic level: Knowledge

67. *Correct answer:* **A**

Appropriate medical treatment or surgical interventions can halt further visual loss, but previous vision loss can't be restored. Frequent ophthalmologic monitoring will catch signs of recurrence or complications.

Nursing process step: Implementation
Client needs category: Physiological integrity
Client needs subcategory: Physiological adaptation
Taxonomic level: Comprehension

68. *Correct answer:* **C**

Aids to communication with the elderly include reducing distracting background noise, beginning the conversation with casual topics, sticking to a topic for a while, rephrasing misunderstood sentences, and keeping sentences and questions short.

Nursing process step: Implementation
Client needs category: Physiological integrity
Client needs subcategory: Physiological adaptation
Taxonomic level: Application

69. *Correct answer:* **D**

Presbycusis is sensorineural hearing loss associated with aging. The other types of hearing loss aren't usually associated with aging.

Nursing process step: Assessment
Client needs category: Health promotion and maintenance
Client needs subcategory: Growth and development through the life span
Taxonomic level: Knowledge

70. *Correct answer:* **C**

Before inserting the hearing aid, check to see that it's working. A squealing noise will be heard as the volume is turned higher. The hearing aid should remain in place throughout the day. Always speak directly toward a patient who's hard of hearing.

Nursing process step: Implementation
Client needs category: Physiological integrity
Client needs subcategory: Physiological adaptation
Taxonomic level: Knowledge

71. *Correct answer:* **B**

Elderly patients typically can't hear high-pitched voices and can't distinguish voices in a noisy environment. The use of a properly functioning hearing aid should correct this problem.

Nursing process step: Assessment
Client needs category: Physiological integrity
Client needs subcategory: Reduction of risk potential
Taxonomic level: Knowledge

72. *Correct answer:* **C**

A microphone receives environmental and speech sounds and converts them to electrical signals, which are then amplified. A receiver then converts the signals to sound.

Nursing process step: N/A
Client needs category: N/A
Client needs subcategory: N/A
Taxonomic level: Knowledge

73. *Correct answer:* **A**

Trigeminal neuralgia is a syndrome of excruciating, intermittent, paroxysmal facial pain. It manifests as intense, periodic pain in the lips, gums, teeth, or chin. The other symptoms aren't characteristic of trigeminal neuralgia.

Nursing process step: Assessment
Client needs category: Physiological integrity
Client needs subcategory: Physiological adaptation
Taxonomic level: Comprehension

74. *Correct answer:* **B**

Trigeminal neuralgia is a disorder of one or more branches of the trigeminal or fifth cranial nerve. No other cranial nerves are involved.

> *Nursing process step:* Assessment
> *Client needs category:* Physiological integrity
> *Client needs subcategory:* Physiological adaptation
> *Taxonomic level:* Knowledge

75. *Correct answer:* **C**

The anticonvulsant drugs carbamazepine (Tegretol) and phenytoin (Dilantin) are the first mode of treatment for trigeminal neuralgia, which is followed by minor or major surgical intervention. High-dose corticosteroids aren't helpful in treating this condition.

> *Nursing process step:* Planning
> *Client needs category:* Physiological integrity
> *Client needs subcategory:* Pharmacological and parenteral therapies
> *Taxonomic level:* Application

76. *Correct answer:* **B**

Mild residual trigeminal nerve edema may decrease the patient's ability to blink. Lubricating drops will prevent dry eyes and the possible resultant damage to the cornea. Additional postoperative instructions will include meticulous oral care, careful assessment of the patient's ability to swallow food and drink, and the avoidance of hot foods because of decreased sensation. An eye patch isn't indicated for this patient.

> *Nursing process step:* Planning
> *Client needs category:* Physiological integrity
> *Client needs subcategory:* Reduction of risk potential
> *Taxonomic level:* Application

DIAGNOSTIC TESTS, THERAPIES, AND TREATMENTS

QUESTIONS

1. A patient is complaining of chest pain that's relieved by sitting forward. The nurse knows that, based on the above assessment data, the cause of this pain is most likely due to:

    A. acute cholecystitis.
    B. acute myocardial infarction.
    C. bronchitis.
    D. pericarditis.

2. A patient who develops a pericardial effusion may have dyspnea, hypotension, and neck vein distention. The nurse anticipates that the diagnostic test that should be performed to confirm this diagnosis is:

    A. an arteriogram
    B. a chest X-ray
    C. a Doppler study
    D. an echocardiogram

3. A patient with a history of rheumatic fever is most likely to develop which disorder as an adult?

    A. Endocarditis
    B. Mitral regurgitation
    C. Myocardial infarction
    D. Pulmonary fibrosis

4. The patient who complains of substernal chest pain that radiates to the left arm and is relieved by rest should be considered for which of the following conditions?

    A. Coronary artery disease
    B. Pericarditis
    C. Pleural effusion
    D. Pneumonia

5. A patient with cardiomyopathy would have which finding on a chest X-ray?

    A. Atelectasis
    B. Cardiomegaly
    C. Mediastinal shift
    D. Pulmonary infiltrate

6. A patient in pulmonary edema exhibits the most distinctive sign of:

    A. bright red sputum.
    B. clear sputum.
    C. green sputum.
    D. pink, frothy sputum.

7. A patient who's suspected of having active tuberculosis is placed in respiratory isolation until a definitive diagnosis is confirmed by:

    A. arterial blood gas sampling.
    B. chest X-ray.
    C. sputum culture and sensitivity.
    D. sputum culture for acid-fast bacilli.

8. A patient with a history of chronic obstructive pulmonary disease (COPD) presents with dyspnea and an oxygen saturation of 88%. What amount of oxygen would the nurse expect to be ordered for this patient?

    A. 2 L via nasal cannula
    B. 6 L via nasal cannula
    C. 50% by face mask
    D. 100% by nonrebreather mask

9. A patient who has had a central venous catheter placed is now complaining of acute shortness of breath. The nurse suspects that the patient has:

    A. an asthma attack.
    B. a pleural effusion.
    C. pneumonia.
    D. a pneumothorax.

10. A postoperative vascular surgery patient has acute shortness of breath and anxiety. The nurse suspects the patient may have a pulmonary embolism. Which of the following tests would the patient be given?

    A. A chest X-ray
    B. An arterial blood gas analysis
    C. A resting energy expenditure scan
    D. A ventilation perfusion scan

11. A patient who's suspected of having a urinary tract infection should have which of the following tests taken to confirm the diagnosis?

    A. Urinalysis
    B. Urine sensitivity test
    C. Urine esterase
    D. Urine metabolites

12. A patient with pneumonia should have which of the following tests performed to determine an appropriate antibiotic?

    A. Arterial blood gas
    B. Chest X-ray
    C. Complete blood count
    D. Sputum culture and sensitivity

13. A 30-year-old sexually active female presents with a vaginal discharge and fever. To determine whether the patient has a sexually transmitted disease (STD), which of the following tests should be performed?

    A. Blood chemistries
    B. Cervical swab
    C. Complete blood count (CBC)
    D. Urinalysis

14. Unless otherwise indicated, a patient with a diagnosis of chlamydia can be expected to be treated with:

    A. amoxicillin.
    B. ampicillin.
    C. penicillin.
    D. tetracycline.

15. A patient who's found to have pelvic inflammatory disease secondary to gonorrhea should be treated with:

    A. amoxicillin.
    B. erythromycin.
    C. penicillin.
    D. sulfonamides.

16. The blood test that's performed to determine whether a patient has syphilis is:

    A. androgen levels.
    B. complete blood count (CBC).
    C. C-reactive protein.
    D. rapid plasma reagin (RPR).

17. A 28-year-old male patient complains of right lower quadrant abdominal pain with rebound tenderness, nausea and vomiting for 2 days, and a temperature of 100° F (37.7° C). The nurse suspects that the patient most likely has:

    A. appendicitis.
    B. cholecystitis.
    C. colitis.
    D. diverticulitis.

18. A patient has a history of chronic renal failure and is being treated with continuous ambulatory peritoneal dialysis (CAPD). The dialysate effluent received back one morning is cloudy. The nurse suspects the patient has:

    A. a bowel perforation.
    B. appendicitis.
    C. cholecystitis.
    D. peritonitis.

19. A patient complains of fatigue and weakness and reports heavier-than-usual menses. The nurse might suspect the patient has:

    A. anemia.
    B. gonorrhea.
    C. mononucleosis.
    D. syphilis.

20. A patient on I.V. heparin should have which of the following laboratory values monitored closely to determine whether the therapeutic range is maintained?

    A. Hemoglobin
    B. International normalized ratio (INR)
    C. Partial thromboplastin time (PTT)
    D. Prothrombin time (PT)

21. A patient receiving warfarin (Coumadin) should have which of the following laboratory values closely monitored?

    A. Hemoglobin
    B. Hemoglobin A_{1C}
    C. Partial thromboplastin time (PTT)
    D. Prothombin time (PT)

22. The nurse expects a patient with thrombocytopenia to have a decreased amount of:

    A. lymphocytes.
    B. platelets.
    C. red blood cells.
    D. white blood cells.

23. The nurse understands that a diabetic patient should have which of the following tests performed to determine the effect of therapy?

    A. Complete blood count
    B. Hemoglobin A_{1C}
    C. Potassium
    D. Sedimentation rate

24. A patient who has a potassium level of 6.0 should be treated with:

    A. antacids.
    B. I.V. fluids.
    C. fluid restriction.
    D. sodium polystyrene sulfonate (Kayexalate).

25. Hyperkalemia can be treated with the administration of 50% dextrose and insulin. The 50% dextrose:

    A. causes potassium to be excreted.
    B. causes potassium to move into the cell.
    C. causes potassium to move into the serum.
    D. counteracts the effects of insulin.

26. A patient has been diagnosed with anemia. To determine whether the condition is due to folic acid deficiency, the nurse expects which of the following tests to be performed?

    A. Folic acid level
    B. Hemoglobin
    C. Schilling test
    D. Vitamin B_{12} level

27. A patient receiving beta blockers may have alterations in which of the following laboratory values?

    A. Glucose
    B. Hemoglobin
    C. Potassium
    D. Sodium

28. A patient receiving diuretics should have which of the following laboratory values closely monitored?

    A. Calcium
    B. Glucose
    C. Phosphorus
    D. Potassium

29. A patient with gout would have an elevation in which of the following laboratory values?

    A. Glucose
    B. Potassium
    C. Erythrocyte sedimentation rate
    D. Uric acid

30. The laboratory values typically assessed when a patient is suspected of having pancreatitis are:

    A. amylase and lipase.
    B. blood urea nitrogen and creatinine.
    C. calcium and potassium.
    D. liver function.

31. A patient is receiving I.V. fluids and is diuresing large amounts of urine. Which of the following tests should be performed?

    A. Red blood cell count
    B. Urine culture and sensitivity
    C. Urine specific gravity
    D. White blood cell count

32. A patient who reports vomiting bright red blood should be suspected of having:

    A. gastroesophageal reflux.
    B. hiatal hernia.
    C. lower GI hemorrhage.
    D. upper GI hemorrhage.

33. A patient with a lower GI hemorrhage would have which of the following findings?

    A. Bloody vomitus
    B. Clay-colored stools
    C. Coffee-ground emesis
    D. Heme-positive stools

34. A patient with a history of alcohol abuse would most likely develop which of the following conditions?

    A. Liver abscess
    B. Liver cancer
    C. Liver cirrhosis
    D. Stomach cancer

35. A patient with a history of alcohol abuse is hospitalized and started on I.V. fluids that include which additives?

    A. Calcium and potassium
    B. Glucose and thiamine
    C. Potassium and glucose
    D. Thiamine and folate

36. The patient with liver failure may develop ascites. This is due to:

    A. bleeding disorders.
    B. dehydration.
    C. protein deficiency.
    D. vitamin deficiency.

37. The patient with liver failure may have alterations in which of the following laboratory values?

    A. Blood urea nitrogen and creatinine
    B. Clotting factors
    C. Creatinine kinase
    D. C-reactive protein

38. A patient with hepatic encephalopathy will have an alteration in which of the following laboratory values?

    A. Ammonia
    B. Glucose
    C. Lactate
    D. Uric acid

39. The patient with hepatic encephalopathy is treated with:

    A. antacids.
    B. beta blockers.
    C. calcium gluconate.
    D. lactulose.

40. A patient with suspected cholelithiasis will have which of the following definitive diagnostic tests performed?

    A. Abdominal computed tomography (CT) scan
    B. Abdominal magnetic resonance imaging (MRI)
    C. Abdominal X-ray
    D. Endoscopic retrograde cholangiopancreatography (ERCP)

41. The patient with a warfarin (Coumadin) overdose should receive which of the following medications to reverse warfarin's effects?

   A. Potassium
   B. Protamine sulfate
   C. Vitamin B$_{12}$
   D. Vitamin K

42. The patient who has hepatitis A most likely contracted it through:

   A. blood products.
   B. dairy products.
   C. I.V. drug use.
   D. seafood ingestion.

43. A patient with a history of seizures who is taking prescribed medication for the condition appears to be having another seizure. The nurse anticipates that the doctor will most likely want to:

   A. give an additional dose of seizure medication.
   B. obtain an electroencephalogram (EEG).
   C. obtain an electrocardiogram (ECG).
   D. prepare for surgery.

44. A patient has Parkinson's disease. What type of medication should the nurse expect the patient to be given?

   A. Antibiotics
   B. Antihistamines
   C. Beta blockers
   D. Dopamine replacement

45. A patient who has had a recent viral infection is now complaining of lower extremity weakness. The nurse suspects that the patient may be developing:

   A. bacterial meningitis.
   B. Guillain-Barré syndrome.
   C. myasthenia gravis.
   D. viral meningitis.

46. The treatment for Guillain-Barré syndrome is:

   A. anti-inflammatory medications.
   B. antiviral medications.
   C. emergency surgery.
   D. supportive in nature.

47. A patient with pyelonephritis complains of urgency and burning on urination, fever, chills, and flank pain. The test that can confirm this diagnosis is:

    A. blood chemistry.
    B. blood glucose.
    C. urinalysis.
    D. urine sodium.

48. A patient undergoes peritoneal dialysis instead of hemodialysis. The nurse understands that the preferred treatment is peritoneal dialysis because it:

    A. can better control the amount of fluid loss.
    B. has less effect on blood pressure.
    C. is more effective in getting rid of excess fluid.
    D. is more effective in getting rid of wastes.

SITUATION: *Susan Mitchell is a 40-year-old female who comes to the hospital with a sudden onset of what she describes as "the worst headache of my life."*

Questions 49 and 50 refer to this situation.

49. The nurse anticipates that Ms. Mitchell may have a:

    A. brain tumor.
    B. cerebral aneurysm.
    C. migraine headache.
    D. subarachnoid hemorrhage.

50. To reduce vasospasm and cerebral infarction, the nurse expects Ms. Mitchell to undergo which of the following procedures?

    A. Aneurysm clipping
    B. Administration of pain medication
    C. Placement of a ventriculostomy
    D. Tumor removal

SITUATION: *Louis Whitman is a 44-year-old male who's admitted to the emergency department with complaints of severe left flank pain, fever, nausea, and vomiting. The nurse suspects that Mr. Whitman has a renal calculus.*

Questions 51 to 53 refer to this situation.

51. The nurse expects Mr. Whitman to have which of the following tests performed to confirm the suspected diagnosis?

    A. Abdominal computed tomography scan
    B. Abdominal ultrasound
    C. Arteriogram
    D. Excretory urography

52. Mr. Whitman will be treated with pain medications and I.V. hydration. Which procedure may be performed if the stone is too large?

    A. Catheterization
    B. Diuretic therapy
    C. Hemodialysis
    D. Lithotripsy

53. When Mr. Whitman is being treated for a renal calculus, the nurse anticipates:

    A. checking specific gravity.
    B. collecting a 24-hour urine sample.
    C. measuring urine output.
    D. straining the urine.

SITUATION: *Edith Kinney is a 63-year-old woman who has been diagnosed with heart failure. She has been admitted to the intensive care unit and is exhibiting shortness of breath.*

Questions 54 to 56 refer to this situation.

54. To determine whether Mrs. Kinney is receiving enough oxygen, the nurse could obtain a:

    A. chest X-ray.
    B. complete blood count.
    C. peak flow value.
    D. pulse oximetry reading.

55. Mrs. Kinney would probably be treated with which class of medications?

    A. Antihypertensive medications
    B. Beta blockers
    C. Diuretics
    D. Nonsteroidal anti-inflammatory drugs

56. Mrs. Kinney has a pulmonary artery (PA) catheter. The nurse would expect to see which of the following values?

    A. Decreased central venous pressure and decreased PA pressures
    B. Decreased central venous pressure and increased PA pressures
    C. Increased central venous pressure and decreased PA pressures
    D. Increased central venous pressure and increased PA pressures

SITUATION: *Ben Draper is admitted to the emergency department with acute shortness of breath after eating peanuts. It's determined that he's an asthmatic and is having an asthma attack.*

Questions 57 and 58 refer to this situation.

57. The best medication to administer to Mr. Draper would be a:

- **A.** beta agonist.
- **B.** beta blocker.
- **C.** calcium channel blocker.
- **D.** vasodilator.

58. To determine how well Mr. Draper's treatment for an acute asthma attack is working, which of the following tests is frequently performed?

- **A.** Blood chemistries
- **B.** Chest X-ray
- **C.** Complete blood count
- **D.** Peak flow

SITUATION: *Carly Watson is a 10-year-old girl who visits the clinic complaining of right ear pain and fever. Carly's mother suspects that she has an ear infection.*

Questions 59 and 60 refer to this situation.

59. Carly may have otitis media. This can be confirmed by an otoscopic examination. The nurse expects to find:

- **A.** bloody discharge from the ear.
- **B.** cerumen in the ear canal.
- **C.** cloudy or serous fluid behind the tympanic membrane.
- **D.** a pearly gray tympanic membrane.

60. The treatment of choice for acute suppurative otitis media is:

- **A.** amoxicillin.
- **B.** antihistamines.
- **C.** beta agonists.
- **D.** decongestants.

SITUATION: *Ralph Sorenson is a 50-year-old diaphoretic male who presents to the emergency department with angina and shortness of breath.*

Questions 61 to 64 refer to this situation.

61. Which of the following tests should be performed to determine whether Mr. Sorenson is having a myocardial infarction (MI)?

    A. Complete blood count (CBC) and creatinine kinase (CK)
    B. Echocardiogram and CK
    C. Electrocardiogram (ECG) and CK
    D. ECG and echocardiogram

62. The best test to determine the extent of coronary artery occlusion is:

    A. a coronary arteriogram.
    B. an echocardiogram.
    C. an electrocardiogram.
    D. an exercise stress test.

63. It's determined that Mr. Sorenson has indeed had a heart attack. The nurse anticipates that he will receive:

    A. antihistamines.
    B. dopamine sulfate.
    C. nonsteroidal anti-inflammatory drugs.
    D. thrombolytic agents.

64. The nurse is providing discharge teaching to Mr. Sorenson. She explains to him that a patient who has coronary artery disease should receive 80 mg of aspirin daily to reduce:

    A. blood pressure.
    B. oxygen demand.
    C. pain.
    D. vascular inflammation.

SITUATION: *Steven Folger is a postoperative patient who has developed acute renal failure from hypotension.*

Questions 65 to 67 refer to this situation.

65. Which of the following tests would assist in confirming that Mr. Folger has acute renal failure?

    A. Calcium
    B. Creatinine
    C. Cystoscopy
    D. Excretory urography

66. The nurse caring for Mr. Folger should closely monitor his:

    A. blood glucose.
    B. blood pressure.
    C. heart rate.
    D. intake and output.

67. If Mr. Folger's renal failure persists, he may require:

    A. antihypertensive medications.
    B. hemodialysis.
    C. I.V. fluids.
    D. vasoactive medications.

SITUATION: *Lois Springer complains of abrupt onset of chest and back pain and loss of radial pulses.*

    Questions 68 to 70 refer to this situation.

68. The nurse suspects that Mrs. Springer may have:

    A. an acute myocardial infarction (MI).
    B. a cerebrovascular accident (CVA).
    C. a dissecting abdominal aneurysm.
    D. a dissecting thoracic aneurysm.

69. The nurse can expect to send Mrs. Springer for which of the following tests to confirm the diagnosis?

    A. Abdominal X-ray
    B. Aortogram
    C. Cardiac catheterization
    D. Chest X-ray

70. The nurse anticipates that the treatment for Mrs. Springer would most likely be:

    A. administration of pain medication.
    B. close monitoring in the patient's room.
    C. emergency surgery to repair the aneurysm.
    D. placement of a ventriculostomy.

SITUATION: *Janet Morris is 1 day postoperative from coronary artery bypass grafting surgery.*

    Questions 71 to 73 refer to this situation.

71. The nurse understands that a postoperative patient who's maintained on bed rest is at high risk for developing:

    A. angina.
    B. arterial bleeding.
    C. deep vein thrombophlebitis (DVT).
    D. dehiscence of the wound.

72. To prevent DVT, the nurse expects that Mrs. Morris will be treated with:

    A. anticoagulants.
    B. I.V. antibiotics.
    C. I.V. hydration.
    D. pain medication.

73. If Mrs. Morris does develop a deep vein thrombosis, it should be treated with:

    A. antibiotics.
    B. anticoagulants.
    C. anti-inflammatory drugs.
    D. I.V. hydration.

SITUATION: *Frank Ehrlich has been admitted to the emergency department with respiratory depression.*

Questions 74 to 76 refer to this situation.

74. Mr. Ehrlich has an ABG analysis with the following results: pH = 7.28, $Paco_2$ = 50, Pao_2 = 75, HCO_3^- = 24, and Sao_2 = 96%. The nurse's interpretation of these results is:

    A. metabolic acidosis.
    B. metabolic alkalosis.
    C. respiratory acidosis.
    D. respiratory alkalosis.

75. The nurse expects the treatment for respiratory acidosis will be to have Mr. Ehrlich:

    A. dump bicarbonate.
    B. increase his breathing rate.
    C. save bicarbonate.
    D. slow his breathing rate.

76. The nurse understands that the cause of Mr. Ehrlich's disorder is:

    A. exercise.
    B. hyperventilation.
    C. kidney failure.
    D. narcotic overdose.

SITUATION: *Molly Peterman is a 25-year-old woman who has been admitted to the emergency department with severe vomiting.*

    Questions 77 to 79 refer to this situation.

77. Ms. Peterman's arterial blood gas (ABG) values are pH = 7.52, $Paco_2$ = 45, Pao_2 = 80, HCO_3^- = 30, and Sao_2 = 94%. The nurse's interpretation of this ABG analysis is:

    A. metabolic acidosis.
    B. metabolic alkalosis.
    C. respiratory acidosis.
    D. respiratory alkalosis.

78. The nurse understands that a possible cause of Ms. Peterman's disorder is:

    A. aspirin overdose.
    B. diabetic ketoacidosis.
    C. narcotic overdose.
    D. renal failure.

79. The nurse anticipates that the treatment of Ms. Peterman's metabolic alkalosis will be to administer:

    A. acetylcysteine (Mucomyst).
    B. hemodialysis.
    C. insulin.
    D. naloxone (Narcan).

ANSWER SHEET

	A B C D		A B C D		A B C D		A B C D
1	○ ○ ○ ○	21	○ ○ ○ ○	41	○ ○ ○ ○	61	○ ○ ○ ○
2	○ ○ ○ ○	22	○ ○ ○ ○	42	○ ○ ○ ○	62	○ ○ ○ ○
3	○ ○ ○ ○	23	○ ○ ○ ○	43	○ ○ ○ ○	63	○ ○ ○ ○
4	○ ○ ○ ○	24	○ ○ ○ ○	44	○ ○ ○ ○	64	○ ○ ○ ○
5	○ ○ ○ ○	25	○ ○ ○ ○	45	○ ○ ○ ○	65	○ ○ ○ ○
6	○ ○ ○ ○	26	○ ○ ○ ○	46	○ ○ ○ ○	66	○ ○ ○ ○
7	○ ○ ○ ○	27	○ ○ ○ ○	47	○ ○ ○ ○	67	○ ○ ○ ○
8	○ ○ ○ ○	28	○ ○ ○ ○	48	○ ○ ○ ○	68	○ ○ ○ ○
9	○ ○ ○ ○	29	○ ○ ○ ○	49	○ ○ ○ ○	69	○ ○ ○ ○
10	○ ○ ○ ○	30	○ ○ ○ ○	50	○ ○ ○ ○	70	○ ○ ○ ○
11	○ ○ ○ ○	31	○ ○ ○ ○	51	○ ○ ○ ○	71	○ ○ ○ ○
12	○ ○ ○ ○	32	○ ○ ○ ○	52	○ ○ ○ ○	72	○ ○ ○ ○
13	○ ○ ○ ○	33	○ ○ ○ ○	53	○ ○ ○ ○	73	○ ○ ○ ○
14	○ ○ ○ ○	34	○ ○ ○ ○	54	○ ○ ○ ○	74	○ ○ ○ ○
15	○ ○ ○ ○	35	○ ○ ○ ○	55	○ ○ ○ ○	75	○ ○ ○ ○
16	○ ○ ○ ○	36	○ ○ ○ ○	56	○ ○ ○ ○	76	○ ○ ○ ○
17	○ ○ ○ ○	37	○ ○ ○ ○	57	○ ○ ○ ○	77	○ ○ ○ ○
18	○ ○ ○ ○	38	○ ○ ○ ○	58	○ ○ ○ ○	78	○ ○ ○ ○
19	○ ○ ○ ○	39	○ ○ ○ ○	59	○ ○ ○ ○	79	○ ○ ○ ○
20	○ ○ ○ ○	40	○ ○ ○ ○	60	○ ○ ○ ○		

ANSWERS AND RATIONALES

1. *Correct answer:* **D**

Inflammation of the pericardial sac causes pain that's much like angina, but that's relieved when sitting forward. Sitting forward moves the pericardial sac away from the epicardium and relieves the pain. Chest pain that occurs in the other conditions is not relieved by sitting forward.

> *Nursing process step:* Analysis
> *Client needs category:* Physiological integrity
> *Client needs subcategory:* Reduction of risk potential
> *Taxonomic level:* Analysis

2. *Correct answer:* **D**

Echocardiograms can diagnose the presence of a pericardial effusion. The other tests can't diagnose this condition.

> *Nursing process step:* Assessment
> *Client needs category:* Physiological integrity
> *Client needs subcategory:* Reduction of risk potential
> *Taxonomic level:* Application

3. *Correct answer:* **B**

Severe cases of rheumatic fever can produce a carditis that can result in valvular disorders. The most common of these valvular disorders is stiffening or swelling of the mitral leaflet leading to mitral regurgitation. The other disorders don't result from rheumatic fever.

> *Nursing process step:* Analysis
> *Client needs category:* Physiological integrity
> *Client needs subcategory:* Physiological adaptation
> *Taxonomic level:* Knowledge

4. *Correct answer:* **A**

The patient with coronary artery disease (CAD) has plaque buildup in the coronary arteries. Chest pain results both from the blockage of blood flow to the area distal to the blockage, and from the tissue demanding more oxygen than is being delivered to it. Although pericarditis symptoms may include pericardial pain as in CAD, the pain usually worsens with deep inspiration and eases when the patient sits up or leans forward. Patients with pleural effusions generally complain of dyspnea and any complaint of pleuritic chest pain is not relieved by rest. The five cardinal signs of pneumonia are coughing, sputum production, pleuritic chest pain, shaking, chills, and fever.

Nursing process step: Analysis
Client needs category: Physiological integrity
Client needs subcategory: Reduction of risk potential
Taxonomic level: Application

5. *Correct answer:* **B**

Cardiomyopathy usually causes the heart chambers to dilate, which would be noted on a chest X-ray as cardiomegaly. Atelectasis is not associated with cardiomyopathy. A mediastinal shift is usually indicative of atelectasis. Pulmonary infiltrates usually indicate pneumonia.

Nursing process step: Assessment
Client needs category: Physiological integrity
Client needs subcategory: Reduction of risk potential
Taxonomic level: Application

6. *Correct answer:* **D**

A patient with pulmonary edema has an increased hydrostatic pressure in the capillary-alveolar bed, which increases sputum production and produces pink, frothy sputum. Bright red sputum generally indicates bleeding. Clear sputum is normal, and green sputum generally indicates infection.

Nursing process step: Analysis
Client needs category: Physiological integrity
Client needs subcategory: Reduction of risk potential
Taxonomic level: Knowledge

7. *Correct answer:* **D**

Sputum culture for acid-fast bacilli gives a definitive diagnosis for tuberculosis when positive. The other options don't give a definitive diagnosis for tuberculosis.

Nursing process step: Assessment
Client needs category: Physiological integrity
Client needs subcategory: Reduction of risk potential
Taxonomic level: Application

8. *Correct answer:* **A**

A patient with COPD breathes by a hypoxic drive that requires a higher partial pressure of arterial carbon dioxide ($Paco_2$) level. If too much oxygen is delivered, the patient may not be able to breathe spontaneously and may require intubation.

DIAGNOSTIC TESTS, THERAPIES, AND TREATMENTS

Nursing process step: Assessment
Client needs category: Physiological integrity
Client needs subcategory: Physiological adaptation
Taxonomic level: Application

9. *Correct answer:* **D**

The patient most likely has developed a pneumothorax from a punctured lung lobe caused by a needle used in the central line insertion. Shortness of breath from a pleural effusion and pneumonia usually has a gradual onset. An asthma attack is sudden, but the recent central line placement makes a pneumothorax first choice.

Nursing process step: Analysis
Client needs category: Physiological integrity
Client needs subcategory: Physiological adaptation
Taxonomic level: Knowledge

10. *Correct answer:* **D**

A patient with a pulmonary embolism would have a ventilation perfusion scan result with a moderate-to-high probability. A chest X-ray rules out other possible causes. The other tests don't provide appropriate diagnostic information.

Nursing process step: Assessment
Client needs category: Physiological integrity
Client needs subcategory: Physiological adaptation
Taxonomic level: Knowledge

11. *Correct answer:* **B**

Urinalysis of a clean catch midstream urine with a bacterial count above 100,000 µl confirms the diagnosis. Urine sensitivity testing determines appropriate therapy. The other urine tests aren't used to diagnose infection.

Nursing process step: Assessment
Client needs category: Physiological integrity
Client needs subcategory: Reduction of risk potential
Taxonomic level: Knowledge

12. *Correct answer:* **D**

Sputum culture and sensitivity will identify the organism and the antibiotic to which the organism is sensitive. The other diagnostic tests can't determine the organism's sensitivity to an antibiotic.

Nursing process step: Assessment
Client needs category: Physiological integrity
Client needs subcategory: Reduction of risk potential
Taxonomic level: Knowledge

13. *Correct answer:* **B**

A cervical swab will identify the organism causing the infection. A CBC, blood chemistries, and urinalysis won't determine whether the patient has an STD.

> *Nursing process step:* Assessment
> *Client needs category:* Physiological integrity
> *Client needs subcategory:* Reduction of risk potential
> *Taxonomic level:* Application

14. *Correct answer:* **D**

Tetracycline by mouth for 7 to 21 days is the treatment of choice for chlamydia infection. The other antibiotics are not effective against chlamydia.

> *Nursing process step:* Planning
> *Client needs category:* Physiological integrity
> *Client needs subcategory:* Pharmacological and parenteral therapies
> *Taxonomic level:* Application

15. *Correct answer:* **C**

Penicillin G is the preferred treatment for gonorrhea.

> *Nursing process step:* Planning
> *Client needs category:* Physiological integrity
> *Client needs subcategory:* Pharmacological and parenteral therapies
> *Taxonomic level:* Knowledge

16. *Correct answer:* **D**

RPR detects nonspecific antibodies within 1 to 2 weeks after a primary lesion appears or 4 to 5 weeks after infection begins. The other blood tests can't detect syphilis.

> *Nursing process step:* Assessment
> *Client needs category:* Physiological integrity
> *Client needs subcategory:* Reduction of risk potential
> *Taxonomic level:* Knowledge

17. *Correct answer:* **A**

Appendicitis presents with pain in the right lower quadrant (known as McBurney's point) accompanied by abdominal rigidity, increasing tenderness, and rebound tenderness. Cholecystitis presents with pain in the right upper quadrant that may radiate to the back, between the shoulders, or to the front of the chest. Colitis presents with bloody diarrhea. Diverticulitis presents with recurrent left lower quadrant pain.

Nursing process step: Assessment
Client needs category: Physiological integrity
Client needs subcategory: Reduction of risk potential
Taxonomic level: Analysis

18. *Correct answer:* **D**

A patient on CAPD is at high risk for peritonitis due to the direct entry of fluid into the peritoneal space. If the effluent becomes cloudy, the patient most likely has peritonitis, which can be life-threatening. A bowel perforation, appendicitis, and peritonitis would not be complications of CAPD.

Nursing process step: Analysis
Client needs category: Physiological integrity
Client needs subcategory: Reduction of risk potential
Taxonomic level: Application

19. *Correct answer:* **A**

The patient's symptoms of fatigue and weakness and the extremely heavy menses indicate anemia. They aren't indications of gonorrhea, mononucleosis, or syphilis.

Nursing process step: Assessment
Client needs category: Physiological integrity
Client needs subcategory: Reduction of risk potential
Taxonomic level: Application

20. *Correct answer:* **C**

A PTT is the only test that measures the effects of I.V. heparin therapy. PT is used to monitor response to oral anticoagulant therapy. The INR system is viewed as the best means of standardizing PT measurements. The other test isn't used to monitor response to anticoagulant therapy.

Nursing process step: Analysis
Client needs category: Physiological integrity
Client needs subcategory: Pharmacological and parenteral therapies
Taxonomic level: Knowledge

21. *Correct answer:* **D**

PT and INR are the laboratory values that give specific information regarding the effectiveness of warfarin (Coumadin) therapy. The other tests aren't appropriate.

Nursing process step: Analysis
Client needs category: Physiological integrity
Client needs subcategory: Pharmacological and parenteral therapies
Taxonomic level: Knowledge

22. *Correct answer:* **B**

A patient with thrombocytopenia will have a low platelet count. Decreased amounts of the other blood cells listed don't occur in thrombocytopenia

> *Nursing process step:* Analysis
> *Client needs category:* Physiological integrity
> *Client needs subcategory:* Reduction of risk potential
> *Taxonomic level:* Knowledge

23. *Correct answer:* **B**

A hemoglobin A_{1C} blood test helps determine the effects of long-term treatment for diabetes. This test gives a more accurate picture of the therapy than does a glucose test, which shows only what the glucose measurement is at the time it's taken. The other tests aren't appropriate in this situation.

> *Nursing process step:* Assessment
> *Client needs category:* Physiological integrity
> *Client needs subcategory:* Pharmacological and parenteral therapies
> *Taxonomic level:* Knowledge

24. *Correct answer:* **D**

Sodium polystyrene sulfonate (Kayexalate) is a resin that pulls potassium into the bowel and is excreted with defecation. Antacids, I.V. fluids, and restriction of fluids will not reduce the potassium level.

> *Nursing process step:* Planning
> *Client needs category:* Physiological integrity
> *Client needs subcategory:* Pharmacological and parenteral therapies
> *Taxonomic level:* Analysis

25. *Correct answer:* **D**

The 50% dextrose is given to counteract the effects of insulin. Insulin drives potassium into the cell, thereby lowering the serum potassium levels. The dextrose doesn't directly cause potassium excretion or any movement of potassium.

> *Nursing process step:* Implementation
> *Client needs category:* Physiological integrity
> *Client needs subcategory:* Pharmacological and parenteral therapies
> *Taxonomic level:* Application

26. *Correct answer:* **C**

A Schilling test determines whether anemia is due to folic acid deficiency or vitamin B_{12} deficiency. The other options represent levels, not tests.

Nursing process step: Planning
Client needs category: Physiological integrity
Client needs subcategory: Reduction of risk potential
Taxonomic level: Analysis

27. *Correct answer:* **A**

Glucose levels—especially in patients who have a history of diabetes—should be monitored closely when taking beta blockers, which can alter glucose metabolism. Beta blockers don't produce alterations in the other laboratory values.

Nursing process step: Evaluation
Client needs category: Physiological integrity
Client needs subcategory: Pharmacological and parenteral therapies
Taxonomic level: Application

28. *Correct answer:* **D**

Hypokalemia is the most common and most serious effect of diuretic use, so potassium must be closely monitored. The other options don't require routine monitoring because diuretics are used.

Nursing process step: Evaluation
Client needs category: Physiological integrity
Client needs subcategory: Pharmacological and parenteral therapies
Taxonomic level: Knowledge

29. *Correct answer:* **D**

Gout is due to an increase in uric acid, which deposits in the joints, such as that of the big toe. Elevated glucose, potassium, and erythrocyte sedimentation rates aren't indicative of a diagnosis of gout.

Nursing process step: Assessment
Client needs category: Physiological integrity
Client needs subcategory: Reduction of risk potential
Taxonomic level: Knowledge

30. *Correct answer:* **A**

Amylase and lipase levels are typically elevated when a patient has pancreatitis. The other options don't determine pancreatic enzyme levels, which would indicate the presence of pancreatitis.

Nursing process step: Assessment
Client needs category: Physiological integrity
Client needs subcategory: Reduction of risk potential
Taxonomic level: Knowledge

31. *Correct answer:* **C**

The urine of a patient receiving I.V. fluids becomes more dilute, which will decrease urine specific gravity. Urine culture and sensitivity tests are unaffected by diuresis. The white blood cell and red blood cell counts aren't tests.

> *Nursing process step:* Assessment
> *Client needs category:* Physiological integrity
> *Client needs subcategory:* Reduction of risk potential
> *Taxonomic level:* Application

32. *Correct answer:* **D**

A patient with an active upper GI hemorrhage would most likely be vomiting bright red blood. Gastroesophageal reflux doesn't always cause symptoms, but its most common feature is heartburn. A hiatal hernia usually is associated with feelings of fullness in the chest and no vomiting. A lower GI hemorrhage would produce blood in the stools.

> *Nursing process step:* Assessment
> *Client needs category:* Physiological integrity
> *Client needs subcategory:* Physiological adaptation
> *Taxonomic level:* Application

33. *Correct answer:* **D**

Lower GI hemorrhage would cause the stool to test heme-positive. Clay-colored stools result from cholelithiasis rather than from GI bleeding. Bloody vomitus and coffee-ground emesis are related to upper GI bleeding.

> *Nursing process step:* Assessment
> *Client needs category:* Physiological integrity
> *Client needs subcategory:* Physiological adaptation
> *Taxonomic level:* Application

34. *Correct answer:* **C**

Liver damage — in this case, cirrhosis — results from malnutrition, especially from dietary protein deficiency. Alcohol hasn't been proven to cause cancer.

> *Nursing process step:* Assessment
> *Client needs category:* Physiological integrity
> *Client needs subcategory:* Reduction of risk potential
> *Taxonomic level:* Knowledge

35. *Correct answer:* **D**

Because of the malnutrition associated with alcohol abuse, folate is given for deficiency anemia and thiamine is given to prevent associated seizures. Most I.V.

solutions contain calcium and potassium; these aren't considered additives. Glucose and thiamine or potassium wouldn't be used in this situation.

Nursing process step: Planning
Client needs category: Physiological integrity
Client needs subcategory: Pharmacological and parenteral therapies
Taxonomic level: Analysis

36. *Correct answer:* **C**

Protein deficiency allows fluid to leak out of the vascular system and third space into the tissues and spaces in the body such as the peritoneal space. Bleeding tendencies, dehydration, and vitamin deficiency can occur but don't cause ascites.

Nursing process step: N/A
Client needs category: Physiological integrity
Client needs subcategory: Physiological adaptation
Taxonomic level: Application

37. *Correct answer:* **B**

Because the liver is responsible for making clotting factors, a patient with liver failure would have alterations in the production of clotting factors. The other values wouldn't identify liver failure.

Nursing process step: Assessment
Client needs category: Physiological integrity
Client needs subcategory: Reduction of risk potential
Taxonomic level: Application

38. *Correct answer:* **A**

Ammonia bypasses the liver and goes directly to the brain, where it causes encephalopathy. Alterations in the other levels don't result directly from hepatic encephalopathy.

Nursing process step: Assessment
Client needs category: Physiological integrity
Client needs subcategory: Reduction of risk potential
Taxonomic level: Knowledge

39. *Correct answer:* **D**

Lactulose works by pulling ammonia into the bowel, where it can be excreted in the stool. Antacids elevate gastric pH to reduce pepsin activity and are used to treat ulcer pain. Beta blockers are used to treat disorders such as hypertension and angina pectoris. Calcium gluconate is used as an antidote for magnesium toxicity.

Nursing process step: Planning
Client needs category: Physiological integrity
Client needs subcategory: Pharmacological and parenteral therapies
Taxonomic level: Application

40. *Correct answer:* **D**

An ERCP visualizes the biliary tree after insertion of an endoscope into the duodenum. Contrast dye is injected to localize the stones. None of the other choices are definitive tests, but they can assist in locating a tumor, obstruction, or mass.

Nursing process step: Assessment
Client needs category: Physiological integrity
Client needs subcategory: Reduction of risk potential
Taxonomic level: Application

41. *Correct answer:* **D**

Vitamin K can reverse the effects of a warfarin (Coumadin) overdose. Protamine sulfate can reverse the effects of a heparin overdose. Vitamin B_{12} and potassium don't affect a warfarin (Coumadin) overdose.

Nursing process step: Planning
Client needs category: Physiological integrity
Client needs subcategory: Pharmacological and parenteral therapies
Taxonomic level: Knowledge

42. *Correct answer:* **D**

Outbreaks of hepatitis A are typically due to people eating seafood from contaminated water. Hepatitis B is typically contracted through blood products and I.V. drug use. Dairy products don't contribute to the spread of any type of hepatitis.

Nursing process step: Assessment
Client needs category: Physiological integrity
Client needs subcategory: Physiological adaptation
Taxonomic level: Application

43. *Correct answer:* **A**

The patient probably doesn't have enough medication in his system to prevent the seizures and requires additional dosing. An EEG confirms a diagnosis of seizure disorder, but this patient has been diagnosed. ECGs monitor cardiac activity. Surgery wouldn't be performed in a patient known to experience seizures.

Nursing process step: Planning
Client needs category: Physiological integrity
Client needs subcategory: Pharmacological and parenteral therapies
Taxonomic level: Application

44. *Correct answer:* **D**

Levodopa, a dopamine-replacement medication, is most effective in the treatment of Parkinson's disease. The other drugs aren't indicated in the treatment of Parkinson's disease.

> *Nursing process step:* Implementation
> *Client needs category:* Physiological integrity
> *Client needs subcategory:* Pharmacological and parenteral therapies
> *Taxonomic level:* Application

45. *Correct answer:* **B**

Guillain-Barré syndrome is frequently correlated to a previous viral syndrome that becomes active in the nerve tracts, causing paresthesia and weakness. The disease is ascending and progressive; then it descends when recovery begins.

> *Nursing process step:* Analysis
> *Client needs category:* Physiological integrity
> *Client needs subcategory:* Reduction of risk potential
> *Taxonomic level:* Application

46. *Correct answer:* **D**

Typically, treatment for Guillain-Barré syndrome involves supporting the patient through the progression, including intubation when respiratory muscles become involved. Prednisone may reduce the progression of the disease. The other treatments aren't appropriate for this syndrome.

> *Nursing process step:* Planning
> *Client needs category:* Physiological integrity
> *Client needs subcategory:* Pharmacological and parenteral therapies
> *Taxonomic level:* Knowledge

47. *Correct answer:* **C**

A urinalysis showing greater than 100,000 organisms/µl of urine and the presence of leukocytes would confirm a diagnosis of pyelonephritis. A blood chemistry would indicate an infection, but it would not help determine the source. Blood glucose and urine sodium would not confirm the diagnosis.

> *Nursing process step:* Assessment
> *Client needs category:* Physiological integrity
> *Client needs subcategory:* Reduction of risk potential
> *Taxonomic level:* Application

48. *Correct answer:* **B**

Peritoneal dialysis is preferred because it has less effect on blood pressure. Hemo-dialysis can remove larger amounts of fluids, and it has a much greater effect on blood pressure. The other two choices are incorrect.

> *Nursing process step:* Evaluation
> *Client needs category:* Physiological integrity
> *Client needs subcategory:* Physiological adaptation
> *Taxonomic level:* Application

49. *Correct answer:* **B**

A cerebral aneurysm rupture will cause sudden, severe headache, nausea, vomit-ing, and possibly an alteration in the level of consciousness. Headaches resulting from a cerebral aneurysm are often described as "the worst headache of my life." Brain tumors, migraines, and subarachnoid hemorrhages may cause severe headache; however, the onset isn't sudden.

> *Nursing process step:* Assessment
> *Client needs category:* Physiological integrity
> *Client needs subcategory:* Physiological adaptation
> *Taxonomic level:* Application

50. *Correct answer:* **A**

Clipping the aneurysm will repair it. Pain medication will relieve the patient's discomfort but not repair the aneurysm. Ventriculostomy isn't performed for an aneurysm. An aneurysm is not a tumor.

> *Nursing process step:* Planning
> *Client needs category:* Physiological integrity
> *Client needs subcategory:* Reduction of risk potential
> *Taxonomic level:* Application

51. *Correct answer:* **D**

Excretory urography will confirm the diagnosis of a renal calculus and will deter-mine the size and location of the calculus. The other tests won't confirm the di-agnosis.

> *Nursing process step:* Planning
> *Client needs category:* Physiological integrity
> *Client needs subcategory:* Reduction of risk potential
> *Taxonomic level:* Knowledge

52. *Correct answer:* **D**

In this type of case — when a stone is believed to be too large to be passed — lithotripsy is performed. This procedure sends sound waves to the area to break the calculus into smaller pieces so they can be passed. Diuretic therapy may be used if the stone is considered small enough to be passed. Hemodialysis and catheterization aren't indicated in the treatment of renal calculi.

> *Nursing process step:* Planning
> *Client needs category:* Physiological integrity
> *Client needs subcategory:* Reduction of risk potential
> *Taxonomic level:* Knowledge

53. *Correct answer:* **D**

To determine whether the stone has been passed, the urine must be strained and the stone must be collected to confirm that it has been passed. Checking specific gravity, collecting a 24-hour urine sample, or measuring urine output won't confirm that the stone has been passed.

> *Nursing process step:* Planning
> *Client needs category:* Safe, effective care environment
> *Client needs subcategory:* Management of care
> *Taxonomic level:* Knowledge

54. *Correct answer:* **D**

Pulse oximetry would show the amount of oxygen the tissues are receiving. The other options would not indicate whether the patient is receiving enough oxygen.

> *Nursing process step:* Planning
> *Client needs category:* Physiological integrity
> *Client needs subcategory:* Physiological adaptation
> *Taxonomic level:* Application

55. *Correct answer:* **C**

Diuretics will increase urinary output, reducing the amount of fluid overload in the lungs and improving oxygenation. The other medications would not necessarily be used to treat shortness of breath related to heart failure.

> *Nursing process step:* Planning
> *Client needs category:* Physiological integrity
> *Client needs subcategory:* Pharmacological and parenteral therapies
> *Taxonomic level:* Application

56. *Correct answer:* **D**

A patient with heart failure is volume overloaded. The pressures in the chambers in the heart would be elevated from the pressure exerted by the increased volume.

Nursing process step: N/A
Client needs category: Physiological integrity
Client needs subcategory: Physiological adaptation
Taxonomic level: Analysis

57. *Correct answer:* **A**

Inhaled beta agonists will cause bronchodilation and improve oxygenation. The other options won't eliminate an asthma attack.

Nursing process step: Planning
Client needs category: Physiological integrity
Client needs subcategory: Pharmacological and parenteral therapies
Taxonomic level: Application

58. *Correct answer:* **D**

A peak flow test requires the patient to take a deep breath and blow out the air through a meter that measures how much air was exhaled. If the bronchodilator has had an effect, an increase in the amount of exhaled air will be noted. The other answers won't determine the effectiveness of treatment.

Nursing process step: Evaluation
Client needs category: Physiological integrity
Client needs subcategory: Physiological adaptation
Taxonomic level: Application

59. *Correct answer:* **C**

Fluid behind the tympanic membrane will appear cloudy, indicating the presence of otitis media. Blood may indicate a more serious finding. A pearly gray tympanic membrane and cerumen are normal findings.

Nursing process step: Assessment
Client needs category: Physiological integrity
Client needs subcategory: Reduction of risk potential
Taxonomic level: Analysis

60. *Correct answer:* **A**

Amoxicillin is the preferred drug for the treatment of acute suppurative otitis media. Decongestants are used to treat acute secretory otitis media. Antihistamines and beta agonists aren't used to treat acute suppurative otitis media.

Nursing process step: Planning
Client needs category: Physiological integrity
Client needs subcategory: Pharmacological and parenteral therapies
Taxonomic level: Knowledge

61. *Correct answer:* **C**

An ECG helps confirm a diagnosis of MI and determines the area of the heart where the MI is occurring. An elevated CK blood level helps confirm myocardial damage. Echocardiograms aren't performed to determine initial diagnosis. A CBC isn't used to diagnose an MI.

> *Nursing process step:* Assessment
> *Client needs category:* Physiological integrity
> *Client needs subcategory:* Reduction of risk potential
> *Taxonomic level:* Analysis

62. *Correct answer:* **A**

A coronary arteriogram will actually visualize the location of the occlusion and the extent of occlusion that's present. The other tests don't determine the extent of coronary artery occlusion.

> *Nursing process step:* Assessment
> *Client needs category:* Physiological integrity
> *Client needs subcategory:* Reduction of risk potential
> *Taxonomic level:* Application

63. *Correct answer:* **D**

A thrombolytic agent will destroy the clot and restore blood flow to the distal region of the heart where flow was restricted. The other medications are not indicated in the treatment of a heart attack.

> *Nursing process step:* Planning
> *Client needs category:* Physiological integrity
> *Client needs subcategory:* Pharmacological and parenteral therapies
> *Taxonomic level:* Application

64. *Correct answer:* **D**

Aspirin has been shown to reduce vascular inflammation, which helps reduce the abrupt occlusion of a coronary artery.

> *Nursing process step:* Implementation
> *Client needs category:* Physiological integrity
> *Client needs subcategory:* Pharmacological and parenteral therapies
> *Taxonomic level:* Knowledge

65. *Correct answer:* **B**

When renal failure occurs, the levels of creatinine rise. Calcium levels, cystoscopy, and excretory urography are associated with the diagnosis of renal calculi.

Nursing process step: Assessment
Client needs category: Physiological integrity
Client needs subcategory: Physiological adaptation
Taxonomic level: Application

66. *Correct answer:* **D**

Intake and output records will give information regarding the effects of therapy and the extent of oliguria due to acute renal failure. Blood pressure and heart rate are routinely monitored. Blood glucose is not routinely monitored in a patient with acute renal failure.

Nursing process step: Implementation
Client needs category: Physiological integrity
Client needs subcategory: Reduction of risk potential
Taxonomic level: Application

67. *Correct answer:* **B**

Hemodialysis will help clear the body of wastes that the body can't remove on its own. The other treatments won't remove body wastes from the blood.

Nursing process step: Implementation
Client needs category: Physiological integrity
Client needs subcategory: Physiological adaptation
Taxonomic level: Knowledge

68. *Correct answer:* **D**

A dissecting thoracic aneurysm may cause loss of radial pulses and severe chest and back pain. An MI typically doesn't cause loss of radial pulses or severe back pain. CVA and dissecting abdominal aneurysm are incorrect responses.

Nursing process step: Assessment
Client needs category: Physiological integrity
Client needs subcategory: Physiological adaptation
Taxonomic level: Analysis

69. *Correct answer:* **B**

An aortogram shows the lumen size and the location of the aneurysm. The other tests don't confirm the diagnosis.

Nursing process step: Implementation
Client needs category: Physiological integrity
Client needs subcategory: Reduction of risk potential
Taxonomic level: Knowledge

70. *Correct answer:* **C**

A dissecting thoracic aneurysm is a surgical emergency. It's essential to stabilize the hemorrhage to prevent exsanguination and death. None of the other options listed will save the patient's life.

> *Nursing process step:* Planning
> *Client needs category:* Physiological integrity
> *Client needs subcategory:* Physiological adaptation
> *Taxonomic level:* Knowledge

71. *Correct answer:* **C**

DVT is the most probable complication for postoperative patients on bed rest. Options A, B, and D aren't likely complications of the postoperative period.

> *Nursing process step:* Planning
> *Client needs category:* Physiological integrity
> *Client needs subcategory:* Reduction of risk potential
> *Taxonomic level:* Application

72. *Correct answer:* **A**

Anticoagulants will prevent deep vein thrombophlebitis if they're started early enough.

> *Nursing process step:* Planning
> *Client needs category:* Physiological integrity
> *Client needs subcategory:* Pharmacological and parenteral therapies
> *Taxonomic level:* Application

73. *Correct answer:* **B**

Anticoagulants can be used after the fact for the prevention of further thromboses and the eventual dissolution of the clot. The other answers are incorrect.

> *Nursing process step:* Planning
> *Client needs category:* Physiological integrity
> *Client needs subcategory:* Pharmacological and parenteral therapies
> *Taxonomic level:* Application

74. *Correct answer:* **C**

Mr. Ehrlich's pH is below 7.35, which makes it acidic, and his $Paco_2$ is above 45, which also makes it acidic. He's in respiratory acidosis because the $Paco_2$ level is controlled by respiration.

Nursing process step: Analysis
Client needs category: Physiological integrity
Client needs subcategory: Reduction of risk potential
Taxonomic level: Analysis

75. *Correct answer:* **B**

An increased respiratory rate would decrease the partial pressure of arterial carbon dioxide ($Paco_2$) level to normal range and correct the acidosis.

> *Nursing process step:* Planning
> *Client needs category:* Physiological integrity
> *Client needs subcategory:* Physiological adaptation
> *Taxonomic level:* Application

76. *Correct answer:* **D**

A narcotic overdose would cause the patient to slow his breathing and retain carbon dioxide. Exercise and hyperventilation can cause respiratory alkalosis. Kidney failure can cause metabolic acidosis.

> *Nursing process step:* Analysis
> *Client needs category:* Physiological integrity
> *Client needs subcategory:* Physiological adaptation
> *Taxonomic level:* Application

77. *Correct answer:* **B**

Ms. Peterman's pH is above 7.45, which makes it alkalotic, and her bicarbonate is high, which also makes it basic. Thus, the diagnosis is metabolic alkalosis.

> *Nursing process step:* Analysis
> *Client needs category:* Physiological integrity
> *Client needs subcategory:* Reduction of risk potential
> *Taxonomic level:* Application

78. *Correct answer:* **D**

Renal failure would impair the excretion of HCO_3^-, causing metabolic alkalosis. Aspirin overdose, diabetic ketoacidosis, and narcotic overdose wouldn't impair the execution of HCO_3^- causing metabolic acidosis.

> *Nursing process step:* Analysis
> *Client needs category:* Physiological integrity
> *Client needs subcategory:* Physiological adaptation
> *Taxonomic level:* Application

79. *Correct answer:* **B**

Hemodialysis would be useful in reducing blood levels of bicarbonate, thereby correcting the disorder. The administration of naloxone (Narcan), acetylcysteine (Mucomyst), and insulin aren't indicated because they don't have any effect on the functioning of the kidneys.

> *Nursing process step:* Planning
> *Client needs category:* Physiological integrity
> *Client needs subcategory:* Physiological adaptation
> *Taxonomic level:* Application

CHAPTER 9

COMPREHENSIVE EXAMINATION

QUESTIONS

1. To determine if tissue underlying the lower lobe of a patient's right lung is filled with fluid, the nurse should use which of the following methods of physical examination?

    A. Auscultation
    B. Inspection
    C. Palpation
    D. Percussion

2. Identify the italicized part of this nurse order: "12/16 Pad side rails *during periods of restlessness and confusion.* P. Jacko, RN."

    A. Content area
    B. Signature
    C. Time element
    D. Verb

3. Which of the following statements regarding the nursing process is true?

    A. It's useful mainly in outpatient settings.
    B. It focuses on the patient, not the nurse.
    C. It progresses in separate, unrelated steps.
    D. It provides the solution to all patient health problems.

4. A patient needs an X-ray study of the gallbladder. The nurse tells the patient that he will be having a:

    A. cholangiography.
    B. cholecystectomy.
    C. cholecystography.
    D. cholecystotomy.

5. The model of care in which one nurse is assigned the overall responsibility for a patient throughout the patient's entire hospitalization period is known as:

    A. case management nursing.
    B. functional nursing.
    C. team nursing.
    D. total patient care nursing.

6. What is the most common form of child abuse?

    A. Child neglect
    B. Emotional abuse
    C. Physical abuse
    D. Sexual abuse

7. A patient diagnosed with a schizophrenic disorder is admitted to the inpatient psychiatric unit. The patient refuses to eat, telling the nurse that spies have put poison in his food. The nurse assesses the patient with the understanding that a false belief to which an individual adheres is:

   A. ambivalence.
   B. anhedonia.
   C. delusion.
   D. psychosis.

8. The nurse-manager tells another nurse to hold down a patient to insert a catheter. The patient says that he doesn't want the catheter inserted. The nurse understands that the patient isn't confused. If the nurse bodily restrains the patient, she's committing:

   A. assault.
   B. battery.
   C. libel.
   D. negligence.

9. A 54-year-old postabdominal surgery patient has been admitted with a wound dehiscence. The nurse is aware that a dehisced wound has:

   A. completely healed.
   B. opened a previously intact suture line.
   C. opened, showing internal organs.
   D. purulent drainage coming from it.

10. Families are defined in many ways. The nuclear family is defined as:

   A. grandparents and grandchildren.
   B. male members of a family.
   C. parent(s) and child or children.
   D. parents, siblings, and first cousins.

11. A quality assurance nurse performs a chart review to determine how many facility patients with surgical incisions are experiencing wound infections. This chart review is an example of what kind of nursing audit?

   A. Concurrent
   B. Outcome
   C. Retrospective
   D. Terminal

12. The nurse understands that the term *presbyopia* means:

A. decreased color discrimination on the blue and green end of the color scale.
B. farsightedness.
C. loss of the ability of the eye to accommodate secondary to loss of the elasticity of the lens.
D. nearsightedness.

13. The nurse is caring for a patient complaining of lower abdominal pain. The doctor diagnoses the patient with a gonadal infection. The nurse knows that the gonads are the:

A. ovaries.
B. ovaries and testes.
C. ovaries, testes, and breasts (male and female).
D. testes.

14. Which of the following choices is an inference rather than a cue?

A. A patient's statement that "it hurts right here"
B. Febrile
C. Hemovac output of 40 ml (1⅜ oz)
D. Blood pressure of 120/80 mm Hg

15. The nurse whose major focus is promoting health screening for deficits and providing support to families would most likely be working in a:

A. hospital.
B. home.
C. long-term facility.
D. school.

16. A nurse is evaluating and judging her colleague's performance on the nursing unit. This is an example of what kind of review?

A. Interdependent
B. Peer
C. Professional
D. Quality

17. The nurse is performing discharge teaching for a newly diagnosed diabetic patient scheduled for a fasting blood glucose test. The nurse explains to the patient that hyperglycemia is defined as a blood glucose level above:

A. 100 mg/dl.
B. 120 mg/dl.
C. 130 mg/dl.
D. 150 mg/dl.

18. Physical assessment findings in the eyes of elderly people may include:

    A. decreased lens thickness.
    B. decreased visual acuity.
    C. lightening of the skin around the orbits.
    D. unequal pupillary light reflex.

19. A patient has been admitted to the intensive care unit with a diagnosis of pericarditis. The nurse understands that patients with pericarditis are treated with:

    A. I.V. morphine sulfate.
    B. I.V. thrombolytic agents.
    C. nonsteroidal anti-inflammatory drugs (NSAIDs).
    D. sublingual nitroglycerin tablets.

20. The 10-year-old child of a Catholic family is hospitalized, and the family requests Communion. This request is an example of what need?

    A. Cultural
    A. Mental
    C. Physical
    D. Spiritual

21. A patient is admitted with a diagnosis of chronic hip pain. By definition, the nurse knows that the patient has had the pain for more than:

    A. 3 months.
    B. 6 months.
    C. 1 year.
    D. 2 years.

22. Which of the following statements is a correctly written actual nursing diagnosis?

    A. Impaired physical mobility as evidenced by decreased range of motion in left shoulder from 180 degrees to 190 degrees of flexion and extension related to left shoulder pain
    B. Ineffective airway clearance related to thickened bronchial secretions as evidenced by adventitious lung sounds over the periphery of the right and left lung fields
    C. Potential for altered nutrition: less than body requirements as evidenced by 15-lb weight loss in 3 weeks
    D. Potential for self-esteem disturbance related to change in body image

23. Many ethnic groups have special needs. The nurse anticipates that an example of a special need Blacks may have is:

    A. foot care.
    B. diet.
    C. hair care.
    D. visiting hours.

24. A patient is admitted to the orthopedic unit following hip replacement surgery. The doctor orders an abduction pillow. An abduction pillow placed after hip replacement surgery prevents:

    A. contracture.
    B. external hip rotation.
    C. footdrop.
    D. internal hip rotation.

25. The nurse observes a patient for signs of distress. Observation is which method of physical examination?

    A. Auscultation
    B. Inspection
    C. Palpation
    D. Percussion

26. The doctor has ordered a medication that's highly irritating to the skin to be given I.M. The nurse uses an I.M. injection method that prevents leakage of the medication into the subcutaneous tissue. This method is known as:

    A. deltoid injection.
    B. intraosseous.
    C. X-track.
    D. Z-track.

27. When performing assessment on a patient, the nurse understands that subjective data are also known as which of the following choices?

    A. Covert data
    B. Inferences
    C. Overt data
    D. Signs

28. Identify the italicized portion of the following potential (risk) nursing diagnosis: "*Potential (risk) for disuse* syndrome related to prescribed bed rest."

A. Inference
B. Etiology (risk factors)
C. Problem statement (diagnostic label)
D. Signs and symptoms (defining characteristics)

29. A patient has a very fast heart rate. To slow the rate down, the doctor asks the patient to hold his breath and bear down. This is known as:

    A. AICD.
    B. cardiopulmonary resuscitation (CPR).
    C. carotid massage.
    D. Valsalva's maneuver.

30. The United States welcomes people of all cultures to live within its boundaries, and the nurse will care for patients of many cultures different from her own. The nurse must first:

    A. examine her own cultural beliefs.
    B. learn more about all cultures.
    C. learn Spanish.
    D. teach the different cultures about America.

31. When performing an admission assessment on a newly admitted patient, the nurse percusses resonance. The nurse knows that resonance heard on percussion is most commonly heard over which organ?

    A. Heart
    B. Liver
    C. Lung
    D. Spleen

32. The nurse is caring for an Italian patient. He's complaining of pain, but he falls asleep right after his complaint and before the nurse can assess his pain. The nurse concludes that:

    A. he probably has a low threshold for pain.
    B. he was faking pain.
    C. someone else gave him medication.
    D. the pain went away.

33. Which of the following statements is true regarding intellect in elderly people?

    A. Approximately 10% suffer from intellectual impairment.
    B. Approximately 20% suffer from intellectual impairment.
    C. 10% of those diagnosed as senile have reversible conditions.
    D. 60% of those diagnosed as senile have reversible conditions.

34. A patient is admitted to the intensive care unit with a diagnosis of closed head injury. The nurse assesses decorticate posturing, which is abnormal:

    A. extension.
    B. flexion.
    C. supination.
    D. pronation.

35. When caring for lower-income patients, the nurse must consider all of the following, *except*:

    A. basic physical needs must be met.
    B. prescriptions may be too costly.
    C. Medicaid will cover all expenses.
    D. transportation to receive health care may be a problem.

36. The nursing order, "12/20 Assist patient to change position. S. Stone, RN," is:

    A. a collaborative nursing intervention.
    B. cumbersome.
    C. not specific enough.
    D. unrealistic.

37. The subjective and objective data that signal the existence of an actual health problem are which of the following parts of an actual nursing diagnosis?

    A. Etiology (risk factors)
    B. Inferences
    C. Problem statement (diagnostic label)
    D. Signs and symptoms (defining characteristics)

38. The nurse is caring for a patient who's admitted to the medical-surgical unit following a cholecystectomy. A T tube drain is in place. The T tube drain is used as a:

    A. biliary drainage tube.
    B. drain after breast surgery.
    C. peritoneal dialysis drain.
    D. small rubber tubing drain.

39. In evaluating the educational process for a community, the nurse would:

    A. continuously evaluate whether the goals and objectives of a program are being met.

    B. monitor the absenteeism rate of a program.

    C. monitor the number of pediatric accident victims at the community hospital emergency department.

    D. perform telephone follow-up of all participants involved.

40. The nurse needs to assess for a thrill of an arteriovenous fistula. The nurse understands that a thrill is:

    A. an audible sound heard with the bell of a stethoscope.

    B. an audible sound heard with the diaphragm of a stethoscope.

    C. a palpable sensation.

    D. usually best assessed by visible inspection.

41. A patient in the community has been diagnosed with coronary artery disease. Which of the following activities would the community health nurse be involved in to promote community-level health promotion activities?

    A. Developing a series of classes on heart disease at a local community hospital

    B. Providing the patient with information on classes on heart disease

    C. Supporting the patient in his attempts to reduce dietary intake of cholesterol

    D. Working with community leaders and citizens to establish heart disease educational programs in the community

42. Which of the following statements about food accurately reflects beliefs about certain foods in various cultural groups?

    A. Asians, Hispanics, and Seventh-Day Adventists won't eat pork.

    B. French patients consider corn to be animal food.

    C. Vegetarian patients generally will include chicken in their diets.

    D. Vietnamese patients won't eat beans.

43. A patient recently developed right-sided weakness that has now resolved. The nurse suspects the patient has had:

    A. a cerebrovascular accident (CVA).

    B. an epileptic event.

    C. a seizure.

    D. a transient ischemic attack (TIA).

44. When caring for a Native American family, the nurse understands that:

    A. Native Americans are incapable of learning how to care for family members.

    B. Native Americans tend to be future oriented.

    C. some Native Americans use herbs in the treatment of illness.

    D. the Native American family unit consists of one parent and the child or children.

45. A patient with an embolic stroke can be treated with:

    A. antibiotics.

    B. anticoagulation agents.

    C. nonsteroidal anti-inflammatory drugs (NSAIDs).

    D. steroids.

46. A patient on bed rest has developed an ulcer that's full thickness and is penetrating the subcutaneous tissue. The nurse documents that this ulcer is in which of the following stages?

    A. Stage 1

    B. Stage 2

    C. Stage 3

    D. Stage 4

47. Any piece of information or data that influences a nurse's decision is:

    A. a cue.

    B. a diagnosis.

    C. an inference.

    D. a judgment.

48. To treat a pressure ulcer, the nurse may use an absorptive seaweed dressing. This type of dressing is known as:

    A. alginate.

    B. gauze.

    C. hydrogel.

    D. transparent film.

49. During the initial assessment of a patient, the nurse percusses dullness. The nurse understands that dullness on percussion can be described as:

    A. flat, with high pitch and short duration.

    B. loud, with high pitch and moderately long duration.

    C. moderate-to-loud, with low pitch and long duration.

    D. soft to moderately loud, with moderate pitch and duration.

50. When considering an individual's cultural background, the term *stereotyping* is defined as:

    A. assuming that patients with similar heritage hold the same customs and beliefs.

    B. expecting all patients to have the same customs that the nurse has.

    C. realizing the effect that socialization has on heritage inconsistency.

    D. realizing all patients have their own beliefs and customs.

51. A patient is admitted to the mental health outpatient agency with a diagnosis of bipolar disorder. The nurse assesses the patient with the understanding that bipolar disorder has two distinct phases, which are:

    A. anxiety and depression.

    B. anxiety and irritability.

    C. mania and anxiety.

    D. mania and depression.

52. During a morning assessment, the nurse percusses tympany. The nurse understands that tympany is a loud, high-pitched, moderately long sound with a drum-like, musical quality that's most commonly heard over the:

    A. heart.

    B. liver.

    C. pancreas.

    D. stomach.

53. To estimate the extent of a burn in an adult patient, the nurse should use which of the following tests?

    A. Lund-Browder

    B. Rule of Nines

    C. Tilt test

    D. Triceps skinfold test

54. The nurse knows that the term *minute volume* refers to the amount of air:

    A. breathed per minute.

    B. inhaled or exhaled during normal breathing.

    C. that can be exhaled after maximum inspiration.

    D. that can be inhaled after normal inspiration.

55. A patient is admitted to the mental health outpatient agency for counseling. The patient complains of being afraid of crowds and won't go into stores until late at night when there are fewer people. The nursing plan of care for this patient is based on the understanding that an irrational fear of something that in reality can cause little if any harm is:

A. panic.
B. phobia.
C. posttraumatic stress.
D. somatoform.

56. Upon a patient's admission to the hospital, the nurse provides the patient with a document written by the American Hospital Association (AHA) that defines patients' rights. This document is called:

A. an informed consent.
B. a legal competence.
C. the Patient's Bill of Rights.
D. the right to know.

57. The nurse elicits rebound tenderness on palpation of the abdomen when the abdomen is:

A. deeply palpated.
B. deeply palpated and quickly released.
C. deeply palpated in two areas at once.
D. lightly palpated.

58. The community health nurse visits the Gomez family every week. They've lived in the United States for six months since moving from Mexico. Although the family income seems adequate, all of the family's money has usually been spent by the end of the month. The nurse knows this is:

A. an indication that Mrs. Gomez doesn't care about her family's needs.
B. an indication that the family doesn't want to discuss money with the nurse.
C. a reflection of the patient's view that the family income is inadequate.
D. a reflection of the patient's culture, values, and time orientation.

59. For potential (risk) nursing diagnoses, nursing interventions will be primarily concerned with which of the following choices?

A. Carrying out the patient's medical regimen
B. Determining the patient's usual patterns
C. Observing for patient for symptoms
D. Providing the patient with health education

60. During a therapeutic encounter, a patient states "I have been very sad." The nurse responds, "So, you say you have been feeling sad?" This is an example of:

    A. empathy.
    B. reflection.
    C. restating.
    D. suggestion.

61. Solutes moving from an area of higher concentration to an area of lower concentration is known as:

    A. active transport.
    B. diffusion.
    C. hydrostatic pressure.
    D. osmosis.

62. Marie and Renaldo Lopez and their three children have recently relocated to the United States from Guatemala. Mrs. Lopez comes to the clinic with acute symptoms of pneumonia. The doctor gives her antibiotics, which she agrees to take. Through her knowledge of Mrs. Lopez's culture, the nurse expects that:

    A. acute care isn't valued.
    B. Mrs. Lopez will bring her children in for immunizations.
    C. Mrs. Lopez won't return to the clinic.
    D. the antibiotics will be taken.

63. A patient is admitted with a diagnosis of left-sided cerebrovascular accident, and assessment reveals right-sided paralysis. The patient is confused and disoriented. The nurse applies a chest or vest restraint to prevent the patient from:

    A. getting out of bed.
    B. moving a part of his body.
    C. pulling out tubes.
    D. rolling over in bed.

64. A patient is complaining of blurred vision following a motor vehicle accident. The nurse knows that the lobe of the brain responsible for vision is the:

    A. motor cortex.
    B. occipital.
    C. parietal.
    D. temporal.

65. An autopsy is a medical examination of a body after death. An autopsy is required by state law if the deceased:

    A. died from human immunodeficiency virus (HIV).
    B. died under suspicious circumstances.
    C. isn't a citizen of the United States.
    D. is under 1 year of age.

66. Identify the italicized portion of the following evaluative statement: "12/20 *Goal met*; oral intake 300 cc more than output, skin turgor good, mucous membranes moist."

    A. Conclusion
    B. Decision
    C. Outcome
    D. Supporting data

67. A patient is admitted to the emergency department with complaints of nausea and vomiting. Arterial blood gas studies show metabolic alkalosis. The nurse understands that alkalosis is a:

    A. pH between 7.15 and 7.33.
    B. pH between 7.35 and 7.45.
    C. pH greater than 7.45.
    D. pH less than 7.35.

68. The most effective means of controlling the transmission of organisms is:

    A. washing one's clothes.
    B. washing one's hair.
    C. washing one's hands.
    D. wearing clean disposable gloves.

69. A patient is going into kidney failure because of rhabdomyolysis. The nurse is understands that rhabdomyolysis is:

    A. abnormal fluid collection in the lungs.
    B. excess water in the cells resulting in cellular swelling.
    C. inflammation of the renal parenchyma, calyces, and pelvis.
    D. skeletal muscle destruction, causing intracellular contents to spill into extracellular fluid.

70. A patient is admitted to the outpatient unit complaining of chest pain. The nurse assesses Erb's point. Erb's point is located in the:

  A. left fifth interspace midclavicular line.
  B. left second intercostal space.
  C. left third to fourth intercostal space.
  D. right second intercostal space.

71. Nu Noi is an 18-year-old exchange student from Thailand from a very poor family. The family he's staying with is very upset because he isn't eating, and they've called a nurse. The nurse would advise the family to:

  A. ask him what he normally eats at home.
  B. grill a steak well done.
  C. make sweet potatoes with sugar.
  D. send him to McDonald's so he has a choice.

72. When giving a back rub to a patient, the nurse should apply:

  A. emollients.
  B. hydrogen peroxide.
  C. rubbing alcohol.
  D. zinc oxide.

73. An Irish immigrant is 1 day postoperative following extensive abdominal surgery. The doctor has ordered 100 mg of meperidine hydrochloride (Demerol) every 4 hours as needed for pain. The nurse notices that the patient hasn't asked for pain medication since surgery. When the nurse assesses the patient, he's stoic and asks to be left alone. The nurse should:

  A. do nothing because he's asked to be left alone.
  B. have one of his family members stay with him.
  C. offer him his pain medication.
  D. report the finding to the doctor.

74. The nurse will most likely need to take initiative in setting priorities when the patient:

  A. has a life-threatening situation.
  B. is demonstrating anger toward the nursing staff.
  C. reads at a sixth-grade level.
  D. uses culturally bound health care practices.

75. Which of the following statements is true regarding goals as opposed to expected outcomes?

  A. Expected outcomes are mutually derived by the nurse and doctor.
  B. Goals are broad statements.
  C. The patient helps determine goals but not expected outcomes.
  D. The doctor must approve goals.

76. The nurse inspects a patient's back and notices small hemorrhagic spots. The nurse documents that the patient has:

    A. extravasation.
    B. osteomalacia.
    C. petechiae.
    D. uremia.

77. The purpose of a Stryker frame is to:

    A. allow horizontal turning of the patient and prevent pressure sores.
    B. allow vertical turning of the patient to increase calcium absorption.
    C. help prevent contractures.
    D. increase respiratory function and decrease hypotension.

78. The nurse assesses a patient with urticaria. The nurse understands that urticaria is another name for:

    A. hives.
    B. a toxin.
    C. a tubercle.
    D. a virus.

79. Which phase of the nursing process establishes the patient's database?

    A. Analysis
    B. Assessment
    C. Evaluation
    D. Implementation

80. The patient tells the nurse that the doctor says he's in remission. The nurse explains that remission means:

    A. the abatement of a disease's symptoms.
    B. a condition that becomes progressively worse and results in death.
    C. the cure of a disease.
    D. the origin and development of a disease.

81. The nurse inspects a patient's pupil size and determines a result of OD = 2 mm and OS = 3 mm. Unequal pupils are known as:

    A. anisocoria.
    B. ataxia.
    C. cataract.
    D. diplopia.

SITUATION: *Ross Simon has been admitted to the hospital with a diagnosis of Addison's disease.*

Questions 82 and 83 refer to this situation.

82. The nurse understands that Addison's disease is a disorder of the:

    A. adrenal gland.
    B. parathyroid gland.
    C. pituitary gland.
    D. thyroid gland.

83. With Addison's disease, the nurse would expect an abnormal laboratory value of:

    A. decreased cortisol level.
    B. decreased thyroid-stimulating hormone.
    C. increased cortisol level.
    D. increased thyroid-stimulating hormone.

SITUATION: *Eddie Joyce has been diagnosed with Graves' disease.*

Questions 84 and 85 refer to this situation.

84. A patient with Graves' disease has a disorder of the:

    A. adrenal gland.
    B. parathyroid gland.
    C. pituitary gland.
    D. thyroid gland.

85. With Graves' disease, the nurse would expect an abnormal laboratory value of:

    A. decreased T_4 (thyroxine).
    B. increased corticotropin.
    C. increased calcium.
    D. increased T_4 (thyroxine).

SITUATION: *Anna Prasad, a 68-year-old woman, is admitted to the general surgical unit for removal of a breast mass malignancy. Chemotherapy will be introduced shortly following the surgery.*

Questions 86 to 95 refer to this situation.

86. Which of the following behaviors or statements by Mrs. Prasad would most support a nursing diagnosis of powerlessness?

A. Her insistence on selecting a specific anesthesiologist
B. Her statement, "No matter what I do, I can't make myself better"
C. Her submission to all recommended therapy
D. Her view that she may die from the illness

87. Mrs. Prasad is Islamic. After the doctor obtains an informed consent from a married Islamic woman, which of the following choices is most important for the nurse to understand?

A. Another Islamic person must co-sign it.
B. It must be explained to her by an Islamic doctor.
C. It must be notarized by Islamic clergy.
D. Her husband must be present when it's signed.

88. In the stress associated with events such as Mrs. Prasad's surgery, which part of the brain mediates the individual's response to the stress?

A. Cerebellum
B. Hypothalamus
C. Medulla
D. Pituitary

89. The nurse discusses guided imagery with Mrs. Prasad before surgery. In guided imagery, the individual focuses on:

A. a mental image.
B. an object such as an icon.
C. a picture or drawing.
D. the person doing the guiding.

90. If cognitive reframing is used as a relaxation or anxiety-reducing technique by the nurse, which characteristic of Mrs. Prasad will the nurse seek to modify?

A. Emotions
B. Memory
C. Perception
D. Sensitivity

91. Which behavior of Mrs. Prasad's would most suggest that guided imagery and cognitive reframing have helped reduce her stress?

A. After the techniques, her pulse and blood pressure are lower.
B. Her family tells the nurse that the patient isn't anxious.
C. She rarely calls the nurses or asks for help in her care.
D. The nurses' notes make no mention of anxiety in the patient.

92. Mrs. Prasad asks the nurse about alternative methods of sexual intimacy. In answering her question, which approach by the nurse is most therapeutic?

    A. Advising her to give herself time to heal before becoming intimate

    B. Giving permission to use different methods of sexual gratification

    C. Referring her to someone with more expertise on the topic

    D. Telling her that postsurgical alternative methods aren't necessary

93. In the presurgical assessment of Mrs. Prasad, which of the following choices would be most appropriate for the nurse to ask Mrs. Prasad when assessing her self-concept?

    A. "Let's talk about what you'd like to do that you haven't done."

    B. "List for me your accomplishments and achievements in life."

    C. "Tell me how this breast surgery will make you feel about yourself."

    D. "What does your husband say when he compliments you?"

94. Before surgery Mrs. Prasad will be given atropine to decrease oral and respiratory secretions. Atropine accomplishes this by:

    A. blocking parasympathetic stimulation.

    B. causing the kidneys to excrete water.

    C. quieting the respiratory center in the brain.

    D. slowing down metabolic activity.

95. Following surgery, Mrs. Prasad will manifest signs of the local adaptation response. Which of the following signs best represents this response?

    A. Fever

    B. Itching

    C. Leukocytosis

    D. Swelling

SITUATION: *Leslie Harper has been diagnosed with Cushing's syndrome.*

Questions 96 and 97 refer to this situation.

96. The nurse understands that Cushing's syndrome is a cluster of abnormalities primarily due to excessive levels of:

    A. adrenocortical hormones.

    B. parathyroid hormones.

    C. pancreatic hormones.

    D. thyroid hormones.

97. With Cushing's syndrome, the nurse would expect an abnormal laboratory value of:

    A. increased antidiuretic hormone.
    B. increased cortisol.
    C. increased growth hormone.
    D. increased thyroid-stimulating hormone.

SITUATION: *The nurse is preparing a nursing plan of care for an elderly patient.*

Questions 98 to 100 refer to this situation.

98. When assessing an elderly patient, the nurse understands that developmental manifestations of elderly patients include:

    A. broadening interests.
    B. decreasing dependency.
    C. decreasing need for affection.
    D. focusing increasingly on self.

99. The nurse understands that one of the major problems of the elderly population in the United States is a:

    A. decreased cost of living.
    B. decreased need for medical care.
    C. poverty rate of 12%.
    D. poverty rate of 20%.

100. The developmental task of elderly people is:

    A. independence versus dependence.
    B. independence versus shame and doubt.
    C. integrity versus despair.
    D. integrity versus independence.

ANSWER SHEET

	A B C D		A B C D		A B C D		A B C D
1	○ ○ ○ ○	26	○ ○ ○ ○	51	○ ○ ○ ○	76	○ ○ ○ ○
2	○ ○ ○ ○	27	○ ○ ○ ○	52	○ ○ ○ ○	77	○ ○ ○ ○
3	○ ○ ○ ○	28	○ ○ ○ ○	53	○ ○ ○ ○	78	○ ○ ○ ○
4	○ ○ ○ ○	29	○ ○ ○ ○	54	○ ○ ○ ○	79	○ ○ ○ ○
5	○ ○ ○ ○	30	○ ○ ○ ○	55	○ ○ ○ ○	80	○ ○ ○ ○
6	○ ○ ○ ○	31	○ ○ ○ ○	56	○ ○ ○ ○	81	○ ○ ○ ○
7	○ ○ ○ ○	32	○ ○ ○ ○	57	○ ○ ○ ○	82	○ ○ ○ ○
8	○ ○ ○ ○	33	○ ○ ○ ○	58	○ ○ ○ ○	83	○ ○ ○ ○
9	○ ○ ○ ○	34	○ ○ ○ ○	59	○ ○ ○ ○	84	○ ○ ○ ○
10	○ ○ ○ ○	35	○ ○ ○ ○	60	○ ○ ○ ○	85	○ ○ ○ ○
11	○ ○ ○ ○	36	○ ○ ○ ○	61	○ ○ ○ ○	86	○ ○ ○ ○
12	○ ○ ○ ○	37	○ ○ ○ ○	62	○ ○ ○ ○	87	○ ○ ○ ○
13	○ ○ ○ ○	38	○ ○ ○ ○	63	○ ○ ○ ○	88	○ ○ ○ ○
14	○ ○ ○ ○	39	○ ○ ○ ○	64	○ ○ ○ ○	89	○ ○ ○ ○
15	○ ○ ○ ○	40	○ ○ ○ ○	65	○ ○ ○ ○	90	○ ○ ○ ○
16	○ ○ ○ ○	41	○ ○ ○ ○	66	○ ○ ○ ○	91	○ ○ ○ ○
17	○ ○ ○ ○	42	○ ○ ○ ○	67	○ ○ ○ ○	92	○ ○ ○ ○
18	○ ○ ○ ○	43	○ ○ ○ ○	68	○ ○ ○ ○	93	○ ○ ○ ○
19	○ ○ ○ ○	44	○ ○ ○ ○	69	○ ○ ○ ○	94	○ ○ ○ ○
20	○ ○ ○ ○	45	○ ○ ○ ○	70	○ ○ ○ ○	95	○ ○ ○ ○
21	○ ○ ○ ○	46	○ ○ ○ ○	71	○ ○ ○ ○	96	○ ○ ○ ○
22	○ ○ ○ ○	47	○ ○ ○ ○	72	○ ○ ○ ○	97	○ ○ ○ ○
23	○ ○ ○ ○	48	○ ○ ○ ○	73	○ ○ ○ ○	98	○ ○ ○ ○
24	○ ○ ○ ○	49	○ ○ ○ ○	74	○ ○ ○ ○	99	○ ○ ○ ○
25	○ ○ ○ ○	50	○ ○ ○ ○	75	○ ○ ○ ○	100	○ ○ ○ ○

ANSWERS AND RATIONALES

1. *Correct answer:* **D**

Percussion is the process of striking a patient's body surface with short, sharp blows of the fingers to determine the size, position, and density of underlying tissue. Auscultation, inspection, or palpation would not help attain this result.

Nursing process step: Assessment
Client needs category: Health promotion and maintenance
Client needs subcategory: Prevention and early detection of disease
Taxonomic level: Comprehension

2. *Correct answer:* **A**

The italicized portion of this nurse order ("during periods of restlessness and confusion") is the content area.

Nursing process step: Planning
Client needs category: Physiological integrity
Client needs subcategory: Reduction of risk potential
Taxonomic level: Comprehension

3. *Correct answer:* **B**

The nursing process is patient-centered, not nurse-centered. It can be used in any setting, and the steps are related. The nursing process can't solve all patient health problems.

Nursing process step: N/A
Client needs category: N/A
Client needs subcategory: N/A
Taxonomic level: Knowledge

4. *Correct answer:* **C**

A cholecystography is an X-ray of the gallbladder. A cholangiography is an X-ray of the bile ducts. A cholecystotomy is an incision and drainage of the gallbladder, and a cholecystectomy is the removal of the gallbladder.

Nursing process step: Implementation
Client needs category: Physiological integrity
Client needs subcategory: Physiological adaptation
Taxonomic level: Knowledge

5. *Correct answer:* **A**

In case management nursing, the patient is assigned to a health care provider who's responsible for coordinating that patient's care as long as the patient is within the provider's service area, such as a hospital or health service. In functional nursing, caregivers are assigned tasks instead of patients. For example, one person would take all the blood pressures. In team nursing, a group of caregivers is assigned a group of patients and the team decides how to care for them. In total patient care nursing, a nurse is assigned a small group of patients and is responsible for all care for those patients.

> *Nursing process step:* N/A
> *Client needs category:* Safe, effective care environment
> *Client needs subcategory:* Management of care
> *Taxonomic level:* Knowledge

6. *Correct answer:* **A**

Child neglect is the most common form of child abuse, followed by physical abuse and then sexual abuse. The incidence of emotional abuse is particularly difficult to quantify.

> *Nursing process step:* N/A
> *Client needs category:* Physiological integrity
> *Client needs subcategory:* Physiological adaptation
> *Taxonomic level:* Knowledge

7. *Correct answer:* **C**

Delusion is a false belief to which an individual adheres. Ambivalence is the simultaneous existence of two opposing feelings, needs, or wishes. Anhedonia is the inability to experience pleasure. Psychosis is a major disturbance in ego functioning.

> *Nursing process step:* Assessment
> *Client needs category:* Physiological integrity
> *Client needs subcategory:* Physiological adaptation
> *Taxonomic level:* Knowledge

8. *Correct answer:* **B**

Battery is the harmful or offensive touching of another's person. Libel is false accusations — written, printed, or typed — that are made with malicious intent. Assault is a threat or attempt to make bodily contact with another person without the other person's consent. Negligence is the omission of an act that a prudent person would have performed.

Nursing process step: Implementation
Client needs category: Safe, effective care environment
Client needs subcategory: Management of care
Taxonomic level: Application

9. *Correct answer:* **B**

Dehiscence of a wound—the partial or total opening of a previously intact suture line—can be preceded by sudden straining, such as coughing and sneezing. An increase in serosanguineous drainage may cause dehiscence. Evisceration is the visibility or protrusion of organs through an incision and requires immediate surgery. Purulent drainage consists of leukocytes, liquefied dead tissue debris, and dead and living bacteria; it's an indication of infection.

Nursing process step: Assessment
Client needs category: Physiological integrity
Client needs subcategory: Physiological adaptation
Taxonomic level: Knowledge

10. *Correct answer:* **C**

A nuclear family is defined as the mother, father, and child or children living in a group. The other options are parts of a nuclear or an extended family.

Nursing process step: N/A
Client needs category: Physiological integrity
Client needs subcategory: Coping and adaptation
Taxonomic level: Knowledge

11. *Correct answer:* **A**

The evaluation of nursing care and patient outcomes while the patients are currently receiving care is called a concurrent nursing audit. The other options aren't types pf audits.

Nursing process step: Evaluation
Client needs category: Safe, effective care environment
Client needs subcategory: Management of care
Taxonomic level: Knowledge

12. *Correct answer:* **C**

Presbyopia is the loss of the ability of the eye to accommodate secondary to loss of the elasticity of the lens. Presbyopia is a normal physiological change with aging. Hyperopia is farsightedness. Myopia is nearsightedness.

Nursing process step: Assessment
Client needs category: Health promotion and maintenance
Client needs subcategory: Growth and development through the life span
Taxonomic level: Knowledge

13. *Correct answer:* **B**

The gonads are the ovaries and testes only.

Nursing process step: Implementation
Client needs category: Physiological integrity
Client needs subcategory: Physiological adaptation
Taxonomic level: Knowledge

14. *Correct answer:* **B**

An inference is the nurse's judgment or interpretation of cues (a cue is any piece of data or information that influences a nurse's decision); in this instance, the nurse infers that the patient is febrile from some unnamed data or information. Data are identified in each of the other options.

Nursing process step: N/A
Client needs category: N/A
Client needs subcategory: N/A
Taxonomic level: Comprehension

15. *Correct answer:* **D**

A school nurse promotes health and screens children for hearing, vision, and scoliosis. Hospital and home care nurses focus on health maintenance. Long-term facility nurses focus on rehabilitation and custodial care.

Nursing process step: N/A
Client needs category: Health promotion and maintenance
Client needs subcategory: Prevention and early detection of disease
Taxonomic level: Application

16. *Correct answer:* **B**

Peer review is an encounter between two persons equal in education, abilities, and qualifications. During the review, one person critically reviews the practices of the other person.

Nursing process step: Evaluation
Client needs category: Safe, effective care environment
Client needs subcategory: Management of care
Taxonomic level: Knowledge

17. *Correct answer:* **B**

Hyperglycemia is defined as a blood glucose level greater than 120 mg/dl. Blood glucose levels of 20 mg/dl, 130 mg/dl, and 150 mg/dl are considered hyperglycemic. A blood glucose level of 100 mg/dl is normal.

Nursing process step: Implementation
Client needs category: Physiological integrity
Client needs subcategory: Reduction of risk potential
Taxonomic level: Knowledge

18. *Correct answer:* **B**

Decreased visual acuity is common in elderly people. Additional assessment findings include increased lens thickness and opacity, darkening of the skin around the orbits, and an equal but slowed pupillary reflex.

Nursing process step: Assessment
Client needs category: Health promotion and maintenance
Client needs subcategory: Growth and development through the life span
Taxonomic level: Knowledge

19. *Correct answer:* **C**

Pericarditis is due to inflammation of the pericardial sac and is treated with NSAIDs. The other medications aren't used to treat pericarditis.

Nursing process step: N/A
Client needs category: Physiological integrity
Client needs subcategory: Pharmacological and parenteral therapies
Taxonomic level: Knowledge

20. *Correct answer:* **D**

Meeting this request is an example of fulfilling a spiritual need. A cultural need is one that meets a particular cultural tradition. A physical need is a physical problem, and a mental need is an emotional or psychological need.

Nursing process step: Analysis
Client needs category: Psychosocial integrity
Client needs subcategory: Coping and adaptation
Taxonomic level: Comprehension

21. *Correct answer:* **B**

Chronic pain is usually defined as pain lasting longer than 6 months. Options A, C, and D are incorrect time periods.

Nursing process step: Assessment
Client needs category: Physiological integrity
Client needs subcategory: Basic care and comfort
Taxonomic level: Knowledge

22. *Correct answer:* **B**

Options C and D are potential diagnoses (not actual nursing diagnoses). In option A, the signs or symptoms (defining characteristics) should follow the etiology (risk factors) and be linked by the phrase "as evidenced by" or "as manifested by."

Nursing process step: Analysis
Client needs category: N/A
Client needs subcategory: N/A
Taxonomic level: Comprehension

23. *Correct answer:* **C**

Blacks may need special hair care because this ethnic group doesn't have oil glands on the scalp as Whites do; thus, they may need oil or lotion massaged on the head to keep hair from breaking. Diet, foot care, and visiting hours aren't any different for Blacks than for any other group.

Nursing process step: Planning
Client needs category: Psychological integrity
Client needs subcategory: Basic care and comfort
Taxonomic level: Comprehension

24. *Correct answer:* **D**

An abductor pillow prevents internal hip rotation. A trochanter roll prevents external hip rotation, and a cradle boot prevents footdrop. Many devices such as a hand roll are available to prevent contractures.

Nursing process step: Implementation
Client needs category: Physiological integrity
Client needs subcategory: Reduction of risk potential
Taxonomic level: Knowledge

25. *Correct answer:* **B**

Inspection is the method of physical examination that uses sight to make deliberate, purposeful observations in a systematic manner with the naked eye or with a lighted instrument such as an otoscope. The other responses don't describe observation.

Nursing process step: Assessment
Client needs category: Health promotion and maintenance
Client needs subcategory: Prevention and early detection of disease
Taxonomic level: Knowledge

26. *Correct answer:* **D**

Z-track injection displaces the skin then releases it after the injection, which prevents seepage of irritating medications into the subcutaneous tissue. The deltoid muscle is a possible site for an I.M. injection. The intraosseous method is an injection into the bone. X-track is an imaginary term.

> *Nursing process step:* Implementation
> *Client needs category:* Physiological integrity
> *Client needs subcategory:* Pharmacological and parenteral therapies
> *Taxonomic level:* Knowledge

27. *Correct answer:* **A**

Subjective data are also known as covert data or symptoms. Overt data and signs can be identified visually. Inferences are conclusions made based on data.

> *Nursing process step:* Assessment
> *Client needs category:* N/A
> *Client needs subcategory:* N/A
> *Taxonomic level:* Knowledge

28. *Correct answer:* **C**

The problem statement (diagnostic label) describes a patient's health problem clearly and concisely in a few words.

> *Nursing process step:* Analysis
> *Client needs category:* Physiological integrity
> *Client needs subcategory:* Reduction of risk potential
> *Taxonomic level:* Comprehension

29. *Correct answer:* **D**

Valsalva's maneuver is used to slow a fast heart rate. An AICD is an implantable internal defibrillator. CPR is used to restore a person's respiration and circulatory status. Carotid massage is also used to slow a fast heart rate by massaging the carotid sinus in the neck.

> *Nursing process step:* Implementation
> *Client needs category:* Physiological integrity
> *Client needs subcategory:* Physiological adaptation
> *Taxonomic level:* Knowledge

30. *Correct answer:* **A**

The nurse must first examine her own cultural beliefs and biases. After the nurse knows her own feelings, she can then begin to understand what is best for her patients.

Nursing process step: Assessment
Client needs category: Psychosocial integrity
Client needs subcategory: Coping and adaptation
Taxonomic level: Application

31. *Correct answer:* C

Resonance is a hollow, moderate-to-loud sound with low pitch and long duration that's heard most commonly over an air-filled tissue such as a normal lung.

Nursing process step: Assessment
Client needs category: Health promotion and maintenance
Client needs subcategory: Prevention and early detection of disease
Taxonomic level: Knowledge

32. *Correct answer:* A

Many people of Italian heritage verbalize discomfort and pain. The pain was real to the patient, and he may need medication when he wakes up.

Nursing process step: Evaluation
Client needs category: Physiological integrity
Client needs subcategory: Pharmacological and parenteral therapies
Taxonomic level: Analysis

33. *Correct answer:* A

About 10% of people over the age of 65 suffer from intellectual impairment. Of those diagnosed as senile, about 20% have reversible conditions, which include drugs and poor nutrition.

Nursing process step: Assessment
Client needs category: Health promotion and maintenance
Client needs subcategory: Growth and development through the life span
Taxonomic level: Knowledge

34. *Correct answer:* B

Decorticate posturing is an abnormal flexion. Decerebrate posturing is an abnormal extension. Both are ominous central nervous system signs. Supination and pronation do not describe decorticate posturing.

Nursing process step: Assessment
Client needs category: Physiological integrity
Client needs subcategory: Physiological adaptation
Taxonomic level: Knowledge

35. *Correct answer:* **C**

Not all lower-income patients are covered by Medicaid; if the patient is covered, Medicaid still may not cover all expenses. Options A, B, and D are correct assumptions when finances are limited.

> *Nursing process step:* Planning
> *Client needs category:* Safe, effective care environment
> *Client needs subcategory:* Management of care
> *Taxonomic level:* Application

36. *Correct answer:* **C**

A complete nursing order has a date, action verb, content area, time element, and signature. In this instance, the time element is missing from the nursing order.

> *Nursing process step:* Planning
> *Client needs category:* Physiological integrity
> *Client needs subcategory:* Reduction of risk potential
> *Taxonomic level:* Comprehension

37. *Correct answer:* **D**

The signs and symptoms (defining characteristics) of an actual nursing diagnosis are the subjective and objective data or information that signal the existence of an actual health problem.

> *Nursing process step:* Analysis
> *Client needs category:* N/A
> *Client needs subcategory:* N/A
> *Taxonomic level:* Knowledge

38. *Correct answer:* **A**

A T tube drain is used as a biliary drainage tube. A Tenckhoff catheter is used as a peritoneal dialysis drain. A Hemovac may be used after breast surgery. A Penrose drain is a small rubber tubing drain.

> *Nursing process step:* Implementation
> *Client needs category:* Physiological integrity
> *Client needs subcategory:* Physiological adaptation
> *Taxonomic level:* Comprehension

39. *Correct answer:* **A**

Process evaluation is necessary throughout a program to ensure that goals and objectives are being met. The other options would not provide sufficient information to determine program success or failure.

Nursing process step: Evaluation
Client needs category: Safe, effective care environment
Client needs subcategory: Management of care
Taxonomic level: Application

40. *Correct answer:* **C**

A thrill is palpated. It can't be assessed with visible inspection and isn't an audible sound that can be heard with the bell or diaphragm of a stethoscope. A bruit is an audible sound heard with the bell of a stethoscope.

Nursing process step: Assessment
Client needs category: Physiological integrity
Client needs subcategory: Physiological adaptation
Taxonomic level: Comprehension

41. *Correct answer:* **D**

Community-level health promotion activities involve the support of community leaders and educators; hence, the nurse would work closely with them to develop a comprehensive approach to the problem of heart disease. All the other responses, although appropriate for this patient, don't reflect community-level activities.

Nursing process step: Implementation
Client needs category: Physiological integrity
Client needs subcategory: Reduction of risk potential
Taxonomic level: Application

42. *Correct answer:* **B**

The French as a group (not individually) don't eat corn. Options A, C, and D are incorrect assumptions.

Nursing process step: N/A
Client needs category: Physiological integrity
Client needs subcategory: Basic care and comfort
Taxonomic level: Knowledge

43. *Correct answer:* **D**

Weakness that resolves is generally due to TIAs, which may be a warning sign of an impending stroke. Typically, CVAs have residual effects. Weakness doesn't usually result from a seizure or an epileptic event.

Nursing process step: Analysis
Client needs category: Physiological integrity
Client needs subcategory: Reduction of risk potential
Taxonomic level: Analysis

44. *Correct answer:* **C**

Native Americans have used herbs for treating illnesses for centuries. Native Americans usually have extended families with many caring people and are oriented very much in today, not to the future.

> *Nursing process step:* Planning
> *Client needs category:* Psychosocial integrity
> *Client needs subcategory:* Coping and adaptation
> *Taxonomic level:* Application

45. *Correct answer:* **B**

A patient with an embolic stroke can be treated with anticoagulation agents to prevent further clot formation and help the existing clot to dissolve. Antibiotics, NSAIDs, and steroids aren't indicated in this condition.

> *Nursing process step:* Planning
> *Client needs category:* Physiological integrity
> *Client needs subcategory:* Pharmacological and parenteral therapies
> *Taxonomic level:* Knowledge

46. *Correct answer:* **C**

A stage 3 ulcer is full thickness involving the subcutaneous tissue. A stage 1 ulcer has a defined area of persistent redness in lightly pigmented skin. A stage 2 ulcer involves partial thickness skin loss. Stage 4 ulcers extend through the skin and exhibit tissue necrosis and muscle or bone involvement.

> *Nursing process step:* Assessment
> *Client needs category:* Physiological integrity
> *Client needs subcategory:* Physiological adaptation
> *Taxonomic level:* Comprehension

47. *Correct answer:* **A**

A cue is any piece of information or data that influences a nurse's decision. An inference or judgment is the nurse's interpretation of cues (any piece of data or information that influences a nurse's decisions). Diagnosis is the analysis of the assessment data to determine the patient's responses that are amendable to nursing intervention.

> *Nursing process step:* Analysis
> *Client needs category:* N/A
> *Client needs subcategory:* N/A
> *Taxonomic level:* Knowledge

48. *Correct answer:* **A**

An absorptive seaweed dressing is an alginate dressing. Gauze is an absorptive cotton or synthetic dressing. Hydrogel is a water-based dressing. Transparent film dressings are clear and nonabsorptive.

> *Nursing process step:* Implementation
> *Client needs category:* Physiological integrity
> *Client needs subcategory:* Physiological adaptation
> *Taxonomic level:* Knowledge

49. *Correct answer:* **D**

Dullness is a thudlike sound, which is soft to moderately loud, is of moderate pitch and duration, and is heard over less dense, mostly fluid-filled matter, such as the liver, heart, and spleen.

> *Nursing process step:* Assessment
> *Client needs category:* Health promotion and maintenance
> *Client needs subcategory:* Prevention and early detection of disease
> *Taxonomic level:* Knowledge

50. *Correct answer:* **A**

Stereotyping is the act of assuming that all members of a culture or ethnic group act alike; thus, option A is the correct answer. Options B, C, and D don't define stereotyping.

> *Nursing process step:* Assessment
> *Client needs category:* Psychosocial integrity
> *Client needs subcategory:* Coping and adaptation
> *Taxonomic level:* Knowledge

51. *Correct answer:* **D**

Bipolar disorder exhibits manic and depressed phases. Anxiety is part of the depressed phase, and irritability is part of the manic phase.

> *Nursing process step:* Assessment
> *Client needs category:* Physiological integrity
> *Client needs subcategory:* Physiological adaptation
> *Taxonomic level:* Knowledge

52. *Correct answer:* **D**

Tympany is produced by air-containing structures, such as the stomach and the bowel.

Nursing process step: Assessment
Client needs category: Health promotion and maintenance
Client needs subcategory: Prevention and early detection of disease
Taxonomic level: Knowledge

53. *Correct answer:* **B**

The Rule of Nines divides the body into 9%, or multiples of 9%, to help estimate the extent of a burn on the body surface. Lund-Browder is similar but is used to estimate burns in children. The triceps skinfold test determines nutritional status by measuring muscle mass. A tilt test is used to diagnose the cause of fainting.

Nursing process step: Assessment
Client needs category: Physiological integrity
Client needs subcategory: Physiological adaptation
Taxonomic level: Knowledge

54. *Correct answer:* **A**

Minute volume refers to the amount of air breathed per minute. The tidal volume is the amount of air inhaled or exhaled during normal breathing. Vital capacity is the amount of air that can be exhaled after maximum inspiration, and inspiratory capacity is the amount of air that can be inhaled after normal inspiration.

Nursing process step: Assessment
Client needs category: Physiological integrity
Client needs subcategory: Physiological adaptation
Taxonomic level: Knowledge

55. *Correct answer:* **B**

An irrational fear of something that in reality can cause little if any harm is a phobia. Panic disorder is unpredictable, paralyzing panic attacks. Somatoform disorder is the transference of an inner conflict onto a body part. Posttraumatic stress disorder is the persistent reexperiencing of a traumatic event.

Nursing process step: Planning
Client needs category: Physiological integrity
Client needs subcategory: Physiological adaptation
Taxonomic level: Knowledge

56. *Correct answer:* **C**

The Patient's Bill of Rights is a document developed by the AHA. Right to know, informed consent, and legal competence are all addressed in the Patient's Bill of Rights.

Nursing process step: Implementation
Client needs category: Safe, effective care environment
Client needs subcategory: Management of care
Taxonomic level: Knowledge

57. *Correct answer:* **B**

Rebound tenderness is due to peritoneal irritation and can be elicited by firm palpation and the quick release of pressure.

Nursing process step: Assessment
Client needs category: Health promotion and maintenance
Client needs subcategory: Prevention and early detection of disease
Taxonomic level: Knowledge

58. *Correct answer:* **D**

The Mexican culture's perception of time focuses on the present and is relatively unconcerned about the future. The Mexican concept of time is a relaxed one.

Nursing process step: Evaluation
Client needs category: Psychosocial integrity
Client needs subcategory: Coping and adaptation
Taxonomic level: Analysis

59. *Correct answer:* **C**

In a potential (risk) nursing diagnosis, the patient isn't currently exhibiting signs or symptoms (defining characteristics) of that nursing diagnosis. However, the nurse must observe the patient for them because the patient has the etiology (risk factors) for that particular nursing diagnosis.

Nursing process step: Planning
Client needs category: N/A
Client needs subcategory: N/A
Taxonomic level: Comprehension

60. *Correct answer:* **C**

Repeating what a patient says is restating. Empathy is experiencing another person's feelings temporarily. Reflection is paraphrasing what a patient says. Suggestion is posing alternatives for a patient.

Nursing process step: Implementation
Client needs category: Psychosocial integrity
Client needs subcategory: Coping and adaptation
Taxonomic level: Knowledge

61. *Correct answer:* **B**

When solutes move from an area of higher concentration to an area of lower concentration, this is known as diffusion. Active transport requires energy to move solutes from an area of lower concentration to an area of higher concentration. Hydrostatic pressure pushes fluids and solutes through a capillary wall. Osmosis occurs when fluid moves passively from an area of high fluid to an area of lower fluid.

> *Nursing process step:* N/A
> *Client needs category:* N/A
> *Client needs subcategory:* N/A
> *Taxonomic level:* Knowledge

62. *Correct answer:* **D**

Guatemalan culture believes in health care for acute illness; thus, the nurse expects that Mrs. Lopez will take her medication to become well. Sense of time and preventive care, such as immunizations, have not been a part of the patient's culture, so reeducation will probably be needed in these areas.

> *Nursing process step:* Evaluation
> *Client needs category:* Psychological integrity
> *Client needs subcategory:* Reduction of risk potential
> *Taxonomic level:* Application

63. *Correct answer:* **A**

Chest or vest restraints are used to prevent a patient from getting out of bed or falling out of the bed or chair. A chest restraint doesn't restrict hand movement, which would prevent the patient from pulling out tubes. Chest restraints should still allow for mobility within the bed, such as rolling onto one side or the other.

> *Nursing process step:* Implementation
> *Client needs category:* Safe, effective care environment
> *Client needs subcategory:* Safety and infection control
> *Taxonomic level:* Comprehensive

64. *Correct answer:* **B**

The occipital lobe of the brain is responsible for vision. The motor cortex is responsible for movement, the parietal lobe is responsible for sensation and body awareness, and the temporal lobe is responsible for hearing, language, comprehension, and recall.

> *Nursing process step:* Assessment
> *Client needs category:* Physiological integrity
> *Client needs subcategory:* Physiological adaptation
> *Taxonomic level:* Knowledge

65. *Correct answer:* **B**

An autopsy is required by law if the death was sudden, occurred under suspicious circumstances, or both. An autopsy also may be performed on a voluntary basis to help determine the cause of death or for research.

> *Nursing process step:* N/A
> *Client needs category:* Safe, effective care environment
> *Client needs subcategory:* Management of care
> *Taxonomic level:* Knowledge

66. *Correct answer:* **A**

The italicized portion of this evaluative statement (either goal met, goal partially met, or goal not met) is the conclusion.

> *Nursing process step:* Evaluation
> *Client needs category:* Physiological integrity
> *Client needs subcategory:* Physiological adaptation
> *Taxonomic level:* Comprehension

67. *Correct answer:* **C**

Alkalosis is a pH greater than 7.45. A pH between 7.35 and 7.45 is normal. A pH less than 7.35 indicates acidosis.

> *Nursing process step:* Assessment
> *Client needs category:* Physiological integrity
> *Client needs subcategory:* Physiological adaptation
> *Taxonomic level:* Comprehension

68. *Correct answer:* **C**

Hand washing is extremely important and is considered the best defense against the transmission of organisms. Maintaining appropriate hygiene measures such as washing hair and clothes is important to prevent the transfer of organisms from the nurse to the patient, but it's most essential that the hands be washed before and after giving care. Wearing clean disposable gloves is necessary when the nurse is likely to handle any body substances, such as blood, urine, and feces; however, the hands still must be washed after removal of the gloves.

> *Nursing process step:* Implementation
> *Client needs category:* Safe, effective care environment
> *Client needs subcategory:* Safety and infection control
> *Taxonomic level:* Knowledge

69. *Correct answer:* **D**

Rhabdomyolysis is a disorder in which skeletal muscle is destroyed, causing intracellular contents to spill into extracellular fluid. Pulmonary edema is abnormal fluid collection in the lungs. Water intoxication is excess water in the cells resulting in cellular swelling. Pyelonephritis is inflammation of the renal parenchyma, calyces, and pelvis.

> *Nursing process step:* Assessment
> *Client needs category:* Physiological integrity
> *Client needs subcategory:* Physiological adaptation
> *Taxonomic level:* Comprehension

70. *Correct answer:* **C**

Erb's point is located in the left third to fourth intercostal space. The mitral area is the left fifth interspace midclavicular line. The pulmonic area is the left second intercostal space. The aortic area is the right second intercostal space.

> *Nursing process step:* Assessment
> *Client needs category:* Physiological integrity
> *Client needs subcategory:* Physiological adaptation
> *Taxonomic level:* Knowledge

71. *Correct answer:* **A**

Asking him what he normally eats at home will open up his likes and dislikes. If a Thai food store is close by, he could choose familiar food and could possibly cook a meal for the host family. The primary meat eaten in Thailand is chicken, and sweet potatoes aren't common there.

> *Nursing process step:* Implementation
> *Client needs category:* Physiological integrity
> *Client needs subcategory:* Basic care and comfort
> *Taxonomic level:* Application

72. *Correct answer:* **A**

Emollients soften and soothe the skin. Keeping the skin clean, dry, and well lubricated helps reduce the development of pressure sores. Rubbing alcohol and hydrogen peroxide are commonly used disinfectants and antiseptics, respectively. Zinc oxide is used to protect and condition sensitive, inflamed areas resulting from contact with diarrhea or enzymatic drainage.

> *Nursing process step:* Implementation
> *Client needs category:* Physiological integrity
> *Client needs subcategory:* Basic care and comfort
> *Taxonomic level:* Knowledge

73. *Correct answer:* **C**

The Irish culture's usual reaction to pain is inexpressive and stoic. Irish people usually don't vocalize that they have pain and will try to hide the pain from family. The nurse should offer the pain medication so that comfort can be extended to the patient.

> *Nursing process step:* Implementation
> *Client needs category:* Physiological integrity
> *Client needs subcategory:* Pharmacological and parenteral therapies
> *Taxonomic level:* Application

74. *Correct answer:* **A**

A patient with a life-threatening situation may not be physiologically able to assist in setting priorities, so the nurse will most likely need to take the initiative in setting priorities for the patient. Patients with the other stated problems should still be able to set priorities, although they may need some direction.

> *Nursing process step:* Planning
> *Client needs category:* Physiological integrity
> *Client needs subcategory:* Physiological adaptation
> *Taxonomic level:* Comprehension

75. *Correct answer:* **B**

Goals are broad statements, whereas expected outcomes are more specific measurable criteria used to evaluate goal achievement. The other options are false statements.

> *Nursing process step:* Planning
> *Client needs category:* N/A
> *Client needs subcategory:* N/A
> *Taxonomic level:* Knowledge

76. *Correct answer:* **C**

Petechiae are small hemorrhagic spots. Extravasation is the leakage of fluid in the interstitial space. Osteomalacia is the softening of bone tissue. Uremia is an excess of urea and other nitrogen products in the blood.

> *Nursing process step:* Assessment
> *Client needs category:* Physiological integrity
> *Client needs subcategory:* Physiological adaptation
> *Taxonomic level:* Knowledge

77. *Correct answer:* **A**

The Stryker frame allows horizontal turning of the patient, which helps prevent pressure sores and increases circulation to prevent blood clots and respiratory problems. Using a Stryker frame will improve a patient's respiratory circulation, but it won't directly improve respiratory function. Vertical positioning of a patient takes place on a tilt table. Passive range-of-motion exercises prevent contractures.

> *Nursing process step:* Implementation
> *Client needs category:* Physiological integrity
> *Client needs subcategory:* Reduction of risk potential
> *Taxonomic level:* Knowledge

78. *Correct answer:* **A**

Hives and urticaria are two names for the same skin lesion. A toxin is a poison. A tubercle is a tiny round nodule produced by the tuberculosis bacillus. A virus is an infectious parasite.

> *Nursing process step:* Assessment
> *Client needs category:* Physiological integrity
> *Client needs subcategory:* Physiological adaptation
> *Taxonomic level:* Knowledge

79. *Correct answer:* **B**

The assessment phase of the nursing process establishes the patient's database. The other options don't establish the patient's database.

> *Nursing process step:* Assessment
> *Client needs category:* N/A
> *Client needs subcategory:* N/A
> *Taxonomic level:* Knowledge

80. *Correct answer:* **A**

Remission is the abatement of a disease's symptoms. Pathogenesis is the origin and development of a disease. A malignant condition becomes progressively worse and results in death.

> *Nursing process step:* Assessment
> *Client needs category:* Physiological integrity
> *Client needs subcategory:* Physiological adaptation
> *Taxonomic level:* Knowledge

81. *Correct answer:* **A**

Unequal pupils are called anisocoria. Ataxia is uncoordinated actions of involuntary muscle use. A cataract is an opacity of the eye's lens. Diplopia is double vision.

Nursing process step: Assessment
Client needs category: Physiological integrity
Client needs subcategory: Physiological adaptation
Taxonomic level: Knowledge

82. *Correct answer:* **A**

The adrenal gland doesn't produce enough cortisol in a patient with Addison's disease. The other glands listed aren't involved in Addison's disease.

Nursing process step: N/A
Client needs category: Physiological integrity
Client needs subcategory: Physiological adaptation
Taxonomic level: Knowledge

83. *Correct answer:* **A**

The adrenal gland doesn't secrete enough cortisol in a patient with Addison's disease; therefore, laboratory values of cortisol are decreased. Cortisol levels are increased in Cushing's disease. Thyroid-stimulating hormone levels aren't affected by Addison's disease.

Nursing process step: Evaluation
Client needs category: Physiological integrity
Client needs subcategory: Physiological adaptation
Taxonomic level: Application

84. *Correct answer:* **D**

A patient with Graves' disease has hyperthyroidism. The other glands listed aren't involved in Graves' disease.

Nursing process step: N/A
Client needs category: Physiological integrity
Client needs subcategory: Physiological adaptation
Taxonomic level: Knowledge

85. *Correct answer:* **D**

A patient with Graves' disease has hypersecretion of T_4 (thyroxine) and T_3 (triiodothyronine); therefore, T_4 levels are elevated. T_4 levels are decreased in patients with hypothyroidism. Calcium and corticotropin levels aren't affected by Graves' disease.

Nursing process step: Evaluation
Client needs category: Physiological integrity
Client needs subcategory: Reduction of risk potential
Taxonomic level: Application

86. *Correct answer:* **B**

Powerlessness is a nursing diagnosis identified in patients with the perception that their own actions won't significantly affect an outcome. Mrs. Prasad's statement reflects this perception. By itself, her willingness to comply with therapy isn't evidence of powerlessness. It's highly appropriate for patients to select the doctors who provide their care and isn't evidence of powerlessness. Mrs. Prasad's view that she may die from the cancer isn't necessarily a manifestation of powerlessness; instead, it may be evidence that she's reached the acceptance phase of the loss process.

> *Nursing process step:* Assessment
> *Client needs category:* Psychosocial integrity
> *Client needs subcategory:* Psychosocial adaptation
> *Taxonomic level:* Application

87. *Correct answer:* **D**

When a married Islamic woman must sign consent for surgery, her husband must be allowed to be present. The other statements are incorrect.

> *Nursing process step:* Planning
> *Client needs category:* Psychosocial integrity
> *Client needs subcategory:* Psychosocial adaptation
> *Taxonomic level:* Application

88. *Correct answer:* **B**

When an individual perceives a stressor, the hypothalamus mediates the neural and endocrine responses. The cerebellum's main function is to regulate balance and equilibrium. The medulla regulates vital functions. The pituitary is the master gland of the body and secretes a number of hormones involved in such activities as fluid and electrolyte regulation and the body's growth.

> *Nursing process step:* Assessment
> *Client needs category:* Physiological integrity
> *Client needs subcategory:* Reduction of risk potential
> *Taxonomic level:* Analysis

89. *Correct answer:* **A**

In guided imagery, an individual creates a mental image on which she focuses.

> *Nursing process step:* Implementation
> *Client needs category:* Psychosocial integrity
> *Client needs subcategory:* Coping and adaptation
> *Taxonomic level:* Knowledge

90. *Correct answer:* **C**

Cognitive reframing, which is also called thought stopping, is a relaxation or anxiety-reducing technique in which the nurse seeks to change or modify a patient's perception or interpretation of an event. There's no attempt to alter the patient's memory, sensitivity, or emotional state.

> *Nursing process step:* Planning
> *Client needs category:* Psychosocial integrity
> *Client needs subcategory:* Coping and adaptation
> *Taxonomic level:* Analysis

91. *Correct answer:* **A**

Anxiety stimulates the release of catecholamines, which elevate the heart rate and blood pressure. Relaxation techniques, such as guided imagery and cognitive reframing, reduce anxiety, reduce catecholamine release, and lower the heart rate and blood pressure. Patients may often have anxiety or other concerns but don't report them or ask for help. Family members are good sources of information about a patient, but they don't supersede the nurse's observations of the patient. The fact that the nurses' notes mention anxiety isn't by itself diagnostic evidence.

> *Nursing process step:* Evaluation
> *Client needs category:* Psychosocial integrity
> *Client needs subcategory:* Coping and adaptation
> *Taxonomic level:* Application

92. *Correct answer:* **B**

It's important for nurses to address a patient's questions regardless of their own sensitivity. In many instances, the patient is seeking approval or permission to receive sexual satisfaction by means other than those to which she's accustomed. As a professional, the nurse can give that approval or permission. Even after major surgery, intimacy isn't altogether prohibited. Referrals may be necessary in some instances, but the nurse can offer help within her scope of practice, which includes discussing sexuality with her patients. Telling the patient that alternative methods aren't necessary may be true, but it may be dismissing the intent of the patient's question, which is aimed at seeking approval.

> *Nursing process step:* Implementation
> *Client needs category:* Health promotion and maintenance
> *Client needs subcategory:* Growth and development through the life span
> *Taxonomic level:* Application

93. *Correct answer:* **C**

Breast surgery can be psychologically traumatic to a woman, especially if she closely links her womanhood and sexuality to her breasts. With this in mind, the nurse must ask proactive questions regarding the patient's feelings about her own breasts. Listing accomplishments or describing ambitions will give the nurse a general impression of the patient's self-concept; at this point, however, the breast-related question is more valuable because of the nature of the illness and the proposed surgery. Eliciting information about the husband's comments doesn't focus on the nurse's assessment of the patient's own perceptions of herself.

> *Nursing process step:* Implementation
> *Client needs category:* Psychosocial integrity
> *Client needs subcategory:* Coping and adaptation
> *Taxonomic level:* Comprehension

94. *Correct answer:* **A**

The parasympathetic autonomic nervous system stimulates oral and respiratory secretions, which can interfere with airway clearance during surgery. Atropine blocks stimulation of the parasympathetic autonomic nervous system, which reduces these secretions.

> *Nursing process step:* Evaluation
> *Client needs category:* Physiological integrity
> *Client needs subcategory:* Pharmacological and parenteral therapies
> *Taxonomic level:* Knowledge

95. *Correct answer:* **D**

The five signs of the local adaptation response are pain, swelling, heat, redness, and changes in function. Leukocytosis (elevated white blood cell count) and fever are more commonly associated with a systemic problem. Itching (or pruritis) is often a late evidence of local healing and is associated with a systemic response to an antigen.

> *Nursing process step:* Evaluation
> *Client needs category:* Physiological integrity
> *Client needs subcategory:* Reduction of risk potential
> *Taxonomic level:* Knowledge

96. *Correct answer:* **A**

Cushing's syndrome is due to excessive levels of adrenocortical hormones. Excessive levels of parathyroid hormone cause hyperparathyroidism; thyroid hormone causes hyperthyroidism; pancreatic hormone causes diabetes mellitus.

Nursing process step: N/A
Client needs category: Physiological integrity
Client needs subcategory: Reduction of risk potential
Taxonomic level: Application

97. *Correct answer:* **B**

The abnormal adrenocortical hormone secreted with Cushing's syndrome is cortisol. Antidiuretic hormone, growth hormone, and thyroid-stimulating hormone levels are not increased in Cushing's syndrome.

Nursing process step: Evaluation
Client needs category: Physiological integrity
Client needs subcategory: Physiological adaptation
Taxonomic level: Knowledge

98. *Correct answer:* **D**

Developmental manifestations of elderly patients include increasing dependency, narrowed interests, concerns focusing more on the self, and an increasing need for affection.

Nursing process step: Assessment
Client needs category: Health promotion and maintenance
Client needs subcategory: Growth and development through the life span
Taxonomic level: Comprehension

99. *Correct answer:* **C**

One of the major problems of the elderly population in the United States is a poverty rate of 12%. Other problems include an increased cost of living, an increased need for medical care, and decreased income.

Nursing process step: N/A
Client needs category: N/A
Client needs subcategory: N/A
Taxonomic level: Knowledge

100. *Correct answer:* **C**

The developmental task of elderly people is integrity versus despair. The other developmental tasks are associated with other age groups.

Nursing process step: Assessment
Client needs category: Health promotion and maintenance
Client needs subcategory: Growth and development through the life span
Taxonomic level: Knowledge

SELECTED REFERENCES

Assessment Made Incredibly Easy. Springhouse, Pa.: Springhouse Corp., 1998.

Berger, K. *Fundamentals of Nursing,* 2nd ed. Stamford, Conn.: Appleton & Lange, 1998.

Black, J., and Matassarin-Jacobs, E. *Medical-Surgical Nursing: Clinical Management for Continuity of Care,* 5th ed. Philadelphia: W.B. Saunders Co., 1997.

Bloom, B.S., et al. *Taxonomy of Educational Objectives: The Classification of Educational Goals,* Handbook 1. New York: David McKay, 1956.

Brooker, C.G. *Human Structure & Function: Nursing Applications in Clinical Practice,* 2nd ed. St. Louis: Mosby–Year Book, Inc., 1998.

Chernecky, C.C., and Berger, B.J. *Laboratory Tests and Diagnostic Procedures,* 2nd ed. Philadelphia: W.B. Saunders Co., 1997.

Clark, R.A., et al. *Wound Care.* Philadelphia: W.B. Saunders Co., 1998.

Cleveland, L., et al. *Nursing Pharmacology and Clinical Management.* Philadelphia: Lippincott, Williams & Wilkins, 1998.

DeLaune, S., and Ladner, P. *Fundamentals of Nursing Standards and Practice.* New York: Delmar Pubs., 1998.

Friede, A., et al. *CDC Prevention Guidelines: A Guide to Action.* Baltimore: Lippincott, Williams & Wilkins, 1997.

Graham-Brown, R. *Dermatology.* St. Louis: Mosby–Year Book, Inc., 1998.

Horton, R., and Parker, L. *Informed Infection Control Practice.* Philadelphia: W.B. Saunders Co., 1997.

Kee, J.L. *Handbook of Laboratory Diagnostic Tests with Nursing Implications,* 3rd ed. Stamford, Conn.: Appelton & Lange, 1997.

Leahy, J.M, and Kizilay, P.E. *Foundations of Nursing Practice: A Nursing Process Approach.* Philadelphia: W.B. Saunders Co., 1998.

Leavelle, D.E., ed. *Mayo Medical Laboratories: Interpretive Handbook.* Rochester, Minn.: Mayo Medical Laboratories, 1997.

Luckmann, J. *Saunders Manual of Nursing Care.* Philadelphia: W.B. Saunders Co., 1997.

Morison, M. *A Colour Guide to the Nursing Management of Wounds,* 2nd ed. St. Louis: Mosby–Year Book, Inc., 1998.

NCLEX-RN Test Plan. National Council of State Boards of Nursing. Chicago: National Council of State Boards of Nursing, 1997.

Pagana, K.D., and Pagana, T.J. *Mosby's Manual of Diagnostic and Laboratory Tests.* St. Louis: Mosby–Year Book, Inc., 1998.

Potter, P.A., and Perry, A.G. *Fundamentals of Nursing: Concepts, Process and Practice,* 4th ed. St. Louis: Mosby–Year Book, Inc. 1997.

Sussman, C., and Bates-Jensen, B.M. *Wound Care: A Collaborative Practice Manual for Physical Therapists and Nurses.* Gaithersburg, Md.: Aspen Pubs., 1998.

Usatine, R., and Moy, R.L. *Skin Surgery: A Practical Guide.* St. Louis: Mosby–Year Book, Inc., 1998.

INDEX

A

Abdomen
  assessment of, 149, 150
  rebound tenderness on palpation of, 290, 313
Abdominal examination, 127, 149
Abdominal muscles, guarding and, 128, 150
Abduction of a joint, 119, 120, 137
Abduction pillow, 284, 305
ABG values. *See* Arterial blood gas values.
Abuse
  child, 280, 301
  depression and, 94, 114
  homicide within families and, 95, 115
  indicator of, 80, 99
  nursing intervention for patient undergoing, 94, 114
  prevention of, 79, 81, 97, 99
  questioning patient directly about, 94, 114
  risk of suicide and, 95, 115
Achilles tendonitis, 123, 143
Adaptation
  local, signs of, 297, 322
  theory, 61, 74
Addison's disease, 295, 319
Admission, discharge planning and, 32, 45
Adolescents, guidelines for dealing with, 83, 102
Adult respiratory distress syndrome, 6, 18
Affective domain, assessment of, 87, 107
Aging, physical changes associated with, 207, 225
Aging population
  chronic condition among, 92, 112
  cognitive impairment in, 94, 114
  decreased visual acuity in, 283, 304
  depression in, 93, 113
  developmental manifestations of, 298, 323
  goal of health care for, 93, 112
  health promotional activities for, 93, 94, 113
  hearing loss in, 220, 240, 241
  intellectual impairment in, 285, 307
  major fears among, 207, 225

Aging population *(continued)*
  poverty rate of, 298, 323
  trends in health care delivery and, 57, 69
Alcohol abuse
  folate and thiamine given to patient with, 250, 267, 268
  liver cirrhosis caused by, 249, 267
  reason for revoking nursing license, 62, 75
Alginate dressing, 288, 311
Alkalosis, 292, 315
Allen's test
  for assessing patency of radial and ulnar arteries, 7, 19
  indication for, 122, 142
Alzheimer's disease
  correcting nutritional status of patient with, 216, 235, 236
  dementia and, 215, 235
  measures to support family and, 216, 236
  treatment of, 216, 235
Ambulatory peritoneal dialysis, continuous, peritonitis and, 247, 264
American Hospital Association, 64, 77
American Nurses Association, 55, 67, 68, 70
American Red Cross, 54, 67
Amoxicillin, 254, 273
Amylase levels, pancreatitis and, 249, 266
ANA. *See* American Nurses Association.
Anemia
  pernicious, 7, 20
  symptoms of, 247, 264
Aneurysm. *See* Cerebral aneurysm; Thoracic aneurysm.
Anisocoria, 294, 318
Anticholinergic agents, 217, 236, 237
Anticoagulation agents
  for preventing deep vein thrombophlebitis, 257, 276
  for treating embolic stroke, 288, 310
Anxiety
  as barrier to learning, 86, 87, 90, 106, 110
  information about blood transfusion and, 162, 184
  mild and moderate, 169, 193